ANTIPOSITIVIST THEORIES OF THE SCIENCES

AF768238

SOCIOLOGY OF THE SCIENCES MONOGRAPHS

Managing Editors:

R. Whitley, *Manchester Business School, University of Manchester*
L. Graham, *Massachusetts Institute of Technology*
H. Nowotny, *European Centre for Social Welfare Training & Research, Vienna*
P. Weingart, *Universität Bielefeld*

Editorial Board:

G. Böhme, *Technische Hochschule, Darmstadt*
N. Elias, *Universität Bielefeld*
Y. Elkana, *The Van Leer Jerusalem Foundation, Jerusalem*
R. Krohn, *McGill University, Montreal*
W. Lepenies, *Freie Universität Berlin*
H. Martins, *St. Antony's College, Oxford*
E. Mendelsohn, *Harvard University, Massachusetts*
H. Rose, *University of Bradford, Bradford, West Yorkshire*
C. Salomon-Bayet, *Centre National de la Recherche Scientifique, Paris*

1983

NORMAN STOCKMAN
University of Aberdeen, Dept. of Sociology

ANTIPOSITIVIST THEORIES OF THE SCIENCES

Critical Rationalism, Critical Theory and Scientific Realism

Springer-Science+Business Media, B.V.

Library of Congress Cataloging in Publication Data

CIP

Stockman, Norman, 1944–
Antipositivist theories of the sciences.

(Sociology of the sciences monographs)
Bibliography: p.
Includes index.
1. Science–Philosophy. 2. Positivism. I. Title.
II. Series.
Q175.S787 1983 501 83-13877
ISBN 978-90-481-8380-7 ISBN 978-94-015-7678-9 (eBook)
DOI 10.1007/978-94-015-7678-9

All Rights Reserved.
© 1983 by Springer Science+Business Media Dordrecht
Originally published by Kluwer Academic Publishers in 1983
Softcover reprint of the hardcover 1st edition 1983
No part of the material protected by this copyright notice may be reproduced or utilized in any form or by any means, electronic or mechanical, including photocopying, recording or by any information storage and retrieval system, without written permission from the copyright owner.

TABLE OF CONTENTS

PART III: ANTIPOSITIVISM IN THE PHILOSOPHY OF THE SOCIAL SCIENCES

PART IV: POSITIVISM, ANTIPOSITIVISM AND IDEOLOGY

PREFACE

The sciences are too important to be left exclusively to scientists, and indeed they have not been. The structure of scientific knowledge, the role of the sciences in society, the appropriate social contexts for the pursuit of scientific inquiry, have long been matters for reflection and debate about the sciences carried on both within academe and outside it. Even within the universities this reflection has not been the property of any single discipline. Philosophy might have been first in the field, but history and the social sciences have also entered the fray. For the latter, new problems came to the fore, since reflection on the sciences is, in the case of the social sciences, necessarily also reflection on themselves as sciences.

Reflection on the natural sciences and self-reflection by the social sciences came to be dominated in the 1960s by the term 'positivism'. At the time when this word had been invented, the sciences were flourishing; their social and material environment had become increasingly favourable to scientific progress, and the sciences were pointing the way to an optimistic future. In the later twentieth century, however, 'positivism' came to be a word used more frequently by those less sure of nineteenth century certainties. In both sociology and philosophy, 'positivism' was now something to be rejected, and, symbolizing the collapse of an earlier consensus, it became itself the shibboleth of a new dissensus, as different groups of reflective thinkers, in rejecting 'positivism', rejected something different, and often rejected each other.

Yet this chaos has not been altogether unproductive. It has raised the level of our awareness of the philosophical problems of the social sciences, and has reminded us of the interrelationships between sociology and philosophy. And this awareness is no less central to the sociology of the sciences than to any other field of the social sciences.

The aim of this book is, therefore, to extract from the various 'positivist disputes' what is of value to continued reflection and self-reflection in both the natural and the social sciences, before a later generation forgetfully but with a certain justified impatience abandons these tortuous disputes to the detritus of intellectual history. To this end, I have adopted as a method a comparative exposition of three 'antipositivist' theories of the sciences, drawing what is constructive from each rather than setting them at each other's throats in polemical combat. Popper's critical rationalism made the first break away from a self-confident 'logical positivism', rejecting its quest for certainty and its inductive method, but also, and more fundamentally, discovering the idea that the rules of scientific method are a social product, an idea which Popper tried to capture with the concept of 'convention'. Critical theory, especially in the form of Habermas' theory of knowledge-guiding interests, being dissatisfied with the apparent arbitrariness of this 'conventionalism', nonetheless also takes the essentially Kantian step of insisting that scientific knowledge is *human* knowledge; thus the rules of scientific method are to be understood in relation to aspects of the knowing human subject. The third antipositivist theory of science, the realist theory, claims that other writers, including Popper and Habermas, have built their theories of knowledge on a mistaken account of scientific method, and that scientific realism provides a different and more adequate account.

Yet I have also tried to carry the debate beyond this constructive exposition. I have interpreted the realist theory of science not, as its proponents suggest, as a more correct account of scientific method *tout court*, but as a theory of method which is appropriate to a more recent phase in the historical development of certain natural sciences. I have therefore sketched out a synthesis of ideas drawn from all three antipositivist theories of the sciences, the core of which is the thesis that rules of scientific method, while not arbitrary conventions, do change historically in relation to changes in both the material and the social conditions of scientific inquiry. The implications of this thesis seem to me to be quite different for the social sciences than for the natural sciences, and to reinforce the 'antinaturalism' of the critical theorists rather than the 'naturalism' of many scientific realists.

Finally, I have briefly related this partly philosophical, partly sociological account of scientific method to recent, more practical, controversies about the possibly 'ideological' role of science in contemporary societies.

ACKNOWLEDGEMENTS

Many people have helped me in various ways over the years in which this book was being put together. Reinhard Kreckel, Peter McCaffery, and Peter Halfpenny read and commented on the whole of an earlier draft. I have learned much from them and enjoyed their encouragement. Discussions with William Outhwaite and Roy Bhaskar have also been a stimulation, as have those with colleagues and generations of students in the Department of Sociology at Aberdeen, and participants in the History and Philosophy of Science Seminar at Aberdeen, especially Andrew Wear. Richard Whitley, as well as aiding me in the production of this book as editor of the series, also gave me an opportunity some years ago to present an early version of these ideas to the B.S.A. Sociology of Science Study Group, when Ted Benton acted as a challenging discussant. I am grateful to them all. I also want to thank Jeanette Thorn for her excellent typing, Aileen Balfour for unstinting aid in preparing the typescript, and Ellen Wilkinson for secretarial assistance.

Special thanks are due to my wife, Tina, who read every word and gave me every encouragement while I was writing this book.

I should also take responsibility for my translations from the original German texts in my quotations from the following works: H. Albert: 1968, *Traktat über kritische Vernunft*, Mohr, Tübingen; J. Habermas: 1970, *Zur Logik der Sozialwissenschaften*, Suhrkamp, Frankfurt/M.; R. Kreckel: 1975, *Soziologisches Denken: eine kritische Einführung*, Leske Verlag, Opladen; K. Lenk (ed.): 1967, *Ideologie*, Luchterhand, Neuwied and Berlin; H. Schnädelbach: 1969, 'Was ist Ideologie?' *Das Argument* **50**; H. Schnädelbach: 1971, *Erfahrung, Begründung und Reflexion: Versuch über den Positivismus*, Suhrkamp, Frankfurt/M.; A. Wellmer: 1967, *Methodologie als Erkenntnistheorie*, Suhrkamp, Frankfurt/M.

Finally, permission to quote from the publications listed below is acknowledged here with gratitude:

Adorno, T. W., Albert, H., Dahrendorf, R., Habermas, J., Pilot, H., Popper, K. R.: 1976, *The Positivist Dispute in German Sociology*, translated by

Glyn Adey and David Frisby, Heinemann Educational Books Ltd., London (German text: Hermann Luchterhand Verlag, Darmstadt).

Albert, H.: 1968, *Traktat über kritische Vernunft*, J. C. B. Mohr (Paul Siebeck), Tübingen.

Benton, T.: 1977, *Philosophical Foundations of the Three Sociologies*, Routledge and Kegan Paul Ltd., London.

Bhaskar, R.: 1978a, *A Realist Theory of Science*, second edition, Harvester Press, Brighton, Sussex.

Bhaskar, R.: 1978b, 'On the possibility of social scientific knowledge and the limits of naturalism', *Journal for the Theory of Social Behaviour* **8**, pp. 1–28.

Habermas, J.: 1971, *Toward a Rational Society: Student Protest, Science, and Politics*, translated by Jeremy J. Shapiro, Heinemann Educational Books Ltd., London (German text: Suhrkamp Verlag, Frankfurt/M.), and Beacon Press (1970), Boston.

Habermas, J.: 1972, *Knowledge and Human Interests*, translated by Jeremy J. Shapiro, Heinemann Educational Books Ltd., London (German text: Suhrkamp Verlag, Frankfurt/M.), and Beacon Press (1971), Boston.

Harré, R.: 1970, *The Principles of Scientific Thinking*, Macmillan, London and Basingstoke, and the University of Chicago Press, Chicago.

Harré, R. and Secord, P.: 1972, *The Explanation of Social Behaviour*, Basil Blackwell Publisher, Oxford.

Keat, R. and Urry, J.: 1975, *Social Theory as Science*, Routledge and Kegan Paul Ltd., London.

Marx, K.: 1975, *Early Writings*, edited by Quintin Hoare, translated by Rodney Livingstone and Gregor Benton, Penguin Marx Library (© Rodney Livingstone, 1974; © Gregor Benton, 1974; © Lawrence & Wishart, 1973, 1971). Quotations from pp. 209, 322, 355 and 421 reprinted by kind permission of Penguin Books Ltd., Harmondsworth, and Random House Inc., New York.

Popper, K.: 1968, *The Logic of Scientific Discovery* (© 1959 Karl R. Popper), Hutchinson, London, and Basic Books, Inc., Publishers, New York.

Schnädelbach, H.: 1971, *Erfahrung, Begründung und Reflexion: Versuch über den Positivismus*, Suhrkamp Verlag, Frankfurt/M.

PART I

THE IDENTIFICATION OF POSITIVISM

CHAPTER 1

INTRODUCTION

How is positivism to be identified? There have been times when positivism was a creed to which its followers were proud to claim adherence. Both in its early development and in many of its later manifestations, the philosophy for which Auguste Comte invented the name 'Positivism' was propounded and preached with undoubted fervour, so much so that it has been thought to merit analysis as a social movement rather than merely a school of philosophy (Simon, 1972). No less strong was the initial enthusiasm of the members of the Vienna Circle and of those influenced by them for their 'Logical Positivism', although there were those who soon became disturbed by the possibility that their use of that term might result in their too close identification with the Comtist tradition in the eyes of their philosophical adversaries, a consideration which prompted the coinage of the later label, 'Logical Empiricism' (Simon, 1972, p. 259). In both of these cases it would have been both possible and plausible to identify positivism by way of the self-identification of positivists.

Nowadays this situation no longer obtains. In these less enlightened times, a positivist is not a nice thing to be, and nobody will own up to being one. The term, however, has clearly not gone out of currency; if anything it has greater currency now than ever before. It has become, far from a form of self-identification, an accusation: successfully to charge another writer with positivism has become almost a knock-down argument.

To draw attention to this state of affairs has become commonplace. Anthony Giddens, for example, remarks that:

> ... the word 'positivist', like the word 'bourgeois', has become more of a derogatory epithet than a useful descriptive concept, and consequently has been largely stripped of whatever agreed meaning it may once have had (1974, p. ix).

Other writers have vied to provide a pithy or witty characterization of this situation. Raymond Williams asserts that the word 'positivist' "becomes a swear-word, by which nobody is swearing" (1976, p. 201); and Martin

Hollis notes that "nowadays the verb 'to be a Positivist' appears to lack a first person singular for its present tense" (1977, p. 42). That 'positivist' can be used as a form of accusation is however nothing new. The situation may not have been so different even at the time when positivism was a self-proclaimed creed. In *Auguste Comte and Positivism*, John Stuart Mill wrote:

Indeed, though the mode of thought expressed by the terms Positive and Positivism is widely spread, the words themselves are, as usual, better known through the enemies of that mode of thinking than through its friends; and more than one thinker who never called himself or his opinion by those appellations, and carefully guarded himself against being confounded with those who did, finds himself, sometimes to his displeasure, though generally by a tolerably correct instinct, classed with Positivists, and assailed as a Positivist (1961, p. 2).

And to take just one example from a rather later period, the hey-day of logical positivism in Britain in the late 1940s, we encounter the spectacle of two Marxists, Alfred Sohn-Rethel and Maurice Cornforth, arraigning each other on the count of most heinous positivism in the pages of *Modern Quarterly* (Sohn-Rethel, 1947/48; Cornforth, 1947/48).

Such being the case, it is no longer possible to identify positivism through the self-identification of its adherents, since it has none. One solution to the problem thus posed, of how the term is to be used to characterize writers and their ideas on the assumption that such writers will not accept that characterization of their ideas, is politely to decline the challenge and abandon the use of the term except in quotation marks (Platt, 1981). This is perhaps by far the most prudent course to take. Braver or more reckless souls, however, will not be satisfied with this and will attempt, as Giddens puts it, "to impose some order upon the flux of different usages" (1974, p. 2). Given the contentious nature of any given application of the terms 'positivist' and 'positivism', it is precisely here that the problems start. For, as will be shown in this introduction, there are a number of different ways in which this task of imposing, or discovering, order can be conceived. It is, after all, a task of concept-formation, a subject on which there is notoriously little unanimity among either philosophers or sociologists. The various ways in which order may be achieved out of the flux of usage have in turn their roots in different philosophical approaches; awareness of this is, in one sense of that difficult word, reflexivity.

One of the simplest and most forthright methods of dealing with 'positivism' is to define it by reference to its origin. This path is taken by W. M. Simon, who defines the scope of *European Positivism in the Nineteenth Century* as follows:

> By 'Positivism' I mean the doctrine founded by Auguste Comte, not only because Comte himself gave the doctrine that name, but because other usages tend to impoverish the language. For an attitude of admiration for the natural sciences and the wish to extend their virtues to other disciplines we have the excellent word 'scientism', while to apply the word 'positivism' to an abstemious, empiricist refusal to philosophize is plausible but wasteful, as well as in danger of being self-stultifying (1972, p. 3).

A number of writers have pointed out that this is a pedantic and unhelpful device (Halfpenny, 1976, pp. 12ff.; Kolakowski, 1972, pp. 60ff.). Comte may have invented the word, but he did not invent many of the component parts of the doctrine that went by that name, even perhaps components such as the law of three stages which are thought of as essentially Comtean (Sklair, 1970, p. 9). Later thinkers who were prepared to represent themselves as positivists nonetheless did not necessarily claim to be following in the foot-steps of Comte; while some logical positivists did acknowledge Comte as their forerunner in some respects, others, such as A. J. Ayer, disclaim any important connection between Comte and logical positivism (Simon, 1972, p. 262). There were elements of Comte's thought which, it would normally be supposed, run directly counter to the spirit of positivism, most notably the Religion of Humanity. And partly because of that, Comte's work is amenable to varying interpretations, with varying emphases being placed on the different elements of his doctrine.

These are weighty substantive objections to an approach, such as Simon's, which simply identifies positivism with Comte and his doctrine. Lying behind them, however, is an argument which is more methodological in nature. As Raymond Williams puts it:

> It is common practice to speak of the 'proper' or 'strict' meaning of a word by reference to its origins. One of the effects of one kind of classical education, especially in conjunction with one version of the defining function of dictionaries, is to produce what can best be called a sacral attitude to words, and corresponding complaints of vulgar contemporary misunderstanding and misuse. The original meanings of words are always interesting. But what is often most interesting is the subsequent variation (1976, p. 18).

The historical variability of a word is an essential part of its meaning, and one that is lost sight of if the original meaning of a term is taken to be its 'real' meaning; this would be a kind of conceptual essentialism.

An alternative method might be imagined which started out from precisely that historical variability. It would involve the empirical observation of the use of the terms 'positivist' and 'positivism', and the attempt to discern regularities in the empirical data thus discovered. At its most empiricist extreme, it would attempt to formulate conditional hypotheses to explain the circumstances under which these terms would be used. It must be said that no example of such an approach has in fact been found, but it is mentioned as a possibility because it is the logically extreme outcome of treating the historical variability of meaning of the words as a question of empirical variation of use. Its failing would be a failure to treat the use of words as a rule-governed activity: whereas the first approach has an overly restrictive criterion of 'correct' and 'incorrect' usage, this empiricist approach would have no criterion at all.

Thus we are led to a third, and in practice a common approach to imposing order on the flux of different usages. It involves the identification of 'central tenets' or 'characteristic theses' of positivism. A number of writers have identified positivism in this way, and Halfpenny has usefully summarized four different versions, those of Kołakowski (1972, pp. 11ff.), von Wright (1971, p. 4), Abbagano (1967) and Giedymin (1975) in Figure 1, to which is here added Halfpenny's own delimitation of the 'central tenets of positivism'. Each of these identifications of positivism according to its central tenets lends itself to being interpreted as a criterion for 'correct' usage. Each recognizes that there is in fact a wide variety of different usages, diagnoses the confusion and argument at cross purposes which can and does result from this variety, and hopes to afford clarity and more fruitful argument by setting out a standard for correct usage which will avoid terminological misunderstanding (Halfpenny, 1976, p. 20).

The main difficulty with such attempts to identify positivism is that any such proposal for 'correct' usage either reverts to the conceptual essentialism of the first approach discussed above, or else has the status of a stipulative definition.[1] Giedymin, who is particularly sensitive to the variety of ways in which positivism is identified by its opponents, is quite explicit that this is the case: having listed his six central doctrines of 'strict positivism', he points out:

TENETS	CENTRAL TENETS OF POSITIVISM ACCORDING TO: Kołakowski	von Wright	Abbagano	Giedymin	Halfpenny
(1) Unity of science or scientific method	X	X	X	X	X
(2) Empiricism in some form	X	X	X	X	X
(3) Science is the only valid knowledge	X		X	X	
(4) Philosophy is identical to the logic of science			X	X	
(5) Science to be used for the benefit of mankind			X	X	
(6) Mathematical physics as the ideal science		X			
(7) Causal explanations as characteristic of science		X			
(8) Sociological relativism with respect to norms				X	

Fig. 1. Tenets of positivism (adapted from Halfpenny, 1976).

Just how many and which of these doctrines (or programmatic rules) have to be on an author's list in order to regard him as a positivist (antipositivist) is an arbitrary terminological decision. Unless such a decision is made, for example by identifying positivism with strict positivism (the complete list . . .), the concepts of positivism and antipositivism are vague. To use them either as if they were not vague or to imply that the acceptance of one or of only some of the doctrines commits one logically to the acceptance of the remaining ones, is to produce confusions (1975, p. 276).

If the terminological decision, or stipulative definition, is really arbitrary, then it cannot matter very much which decision is taken, as long as all concerned who wish to be able to communicate with each other without misunderstanding stick to the same definition. Arguments about whether a philosophical position is 'really' positivist, or 'close enough' to the definition to be called positivist, ought then not to arise. However, the very fact that there is no such agreement, that the word is used by opponents of 'positivism' with a certain degree of intransigence, that the word, as Williams says, has become a swearword, suggests that the proposal of a stipulative definition to which all could agree, a kind of terminological *contrat social*, is an inadequate response to the situation. It is rooted in the nominalist assumption which Karl Popper sums up in the precept:

Never let yourself be goaded into taking seriously problems about words and their meanings. What must be taken seriously are questions of fact, and assertions about facts; theories and hypotheses; the problems they solve; and the problems they raise (1976, p. 19).

As Popper himself came to realize, behind both these apparently deeply opposed doctrines of essentialism and nominalism lie concepts of clarity and precision which project unrealistic and unrealizable standards for the use of language. One does not have to be a believer in Popper's 'third world' to accept that conceptual problems must be dealt with on their own terms, and that clarity and precision are not to be achieved by fiat.

From here it is a short step to a fourth approach to the identification of positivism. In this approach the variety and divergence of usage of the term 'positivist' and the misunderstanding and confusion to which this gives rise are not seen as problems to be overcome by definitional convention but rather as linguistic facts which have to be understood in their cultural and historical development. The assumptions of such an approach are hermeneutic ones. Writers and speakers are located in historical traditions which are essentially linguistic traditions. The interpretation of the texts which they produce is itself the continuation and development of those traditions. In the course of this historical development the vocabulary of the linguistic traditions takes on shifting, modified, even radically changed meanings as the historical context in which it is used itself changes. In this process words may become embedded in quite different traditions from those in which they originally emerged. Writers and speakers locating themselves in different traditions may then find that they share elements of vocabulary which they nonetheless use in quite divergent ways, although a reconstruction of the historical process of divergence may reveal the common origin and shared assumptions of the variety of meanings. Confusion, misunderstanding, and argument at cross purposes may well come to characterize dialogue between members of different traditions.[2]

It is evident that the divergent usage of the term 'positivism' can be interpreted according to this scheme. Since its invention by Comte, the word has been drafted into service in numerous contexts bearing in some cases only a tenuous relationship to that for which it was originally designed. Its possible sphere of application is by no means restricted to philosophical thought about the natural and social sciences and their relationship to social development. Hence, as is the case also with many other words, an encyclopaedic approach to the range of contexts of its use would pose daunting problems of scale. However, it is necessary to raise the question of the point of such an investigation. Gadamer has argued that hermeneutic interpretation always remains bound to the practical task of maintaining or securing mutual understanding between participants in a dialogue (1975, pp. 345 ff.). It is thus strictly necessary only when such mutual understanding does not obtain, just as speakers of different languages only require an interpreter when they do not share a common language. It follows that an investigation into divergent usage of the term 'positivism' based on hermeneutic assumptions can start out from contexts where mutual understanding is not being achieved between participants in an attempted dialogue. As Popper also argues, clarity and precision are not abstract ideals, rather they must be developed *ad hoc*, when misunderstanding arises which needs to be overcome (1976, p. 30).

Since the argument to be developed here takes its starting point in this hermeneutic approach, it is necessary now to be a little more specific. There are no doubt many potential dialogues between writers who draw on or are located in different traditions; it is necessary therefore to select, and that selection in itself situates the present writer in the context of ongoing debates.

The situation which is to be addressed can be sketched out as a triangular debate between three traditions of philosophical thought about the social and the natural sciences and the relationships between them, each of which conceives of itself as an antipositivist tradition. It has already been suggested that there are no self-confessed positivists taking part in such debates; nonetheless, positivism remains a target of criticism, and whatever positive self-identification may be projected by adherents to a school of thought, the negative self-identification of antipositivism has become *de rigueur*. The three antipositivist traditions referred to can be briefly labelled (1) critical rationalism, (2) critical theory, and (3) scientific realism. Rather than initially identifying each of these labels by their doctrines, theories or *problématiques*, it is more convenient to mention names. By critical rationalism is meant that body of thought developed by Karl Popper, who himself was led to use that term to refer to the general approach embodied in his writings, and by other writers who locate themselves in the same tradition. By critical theory is meant that body of thought often referred to as 'the Frankfurt School', of which Horkheimer, Adorno and Marcuse represent key figures of the older generation, and Jürgen Habermas, Alfred Schmidt, and Albrecht Wellmer a younger generation. By scientific realism is meant that body of thought generated, originally in the philosophy of the natural sciences, by such writers as Rom Harré, Mary Hesse, and Roy Bhaskar, and transferred to the social sciences by, among others, Harré, Bhaskar, Russell Keat and John Urry, and Ted Benton.

The choice of these three antipositivist traditions is related to the debates that have taken place and are continuing to develop between them. In the first place there is the debate known as the *Positivismusstreit*, or positivist dispute, in German sociology. In the connotation this term has taken on since the 1960s, this refers to

> . . . a controversy which has raged in social scientific and philosophical circles in Germany since 1961. The immediate origin of the controversy lay in the conference held by the German Sociological Association in Tübingen in 1961 on the logic of the social sciences in which Popper presented his twenty-seven theses on that topic. This was replied to by Adorno and the discussion which followed at that conference is summarized by Dahrendorf. In a different form, the controversy was continued by Habermas and Albert from 1963 onwards and many other writers took up various issues and aspects of the controversy (Adorno *et al.*, 1976, p. ix).

Most contributors to the controversy located themselves within one of the two traditions at issue, critical rationalism or critical theory, while positivism was, as Dahrendorf put it, 'the third man' in the debate, absent but continually treated as present. The critical theorists persisted in referring to the critical rationalists as positivists, the latter strenuously resisted that appellation, and occasionally even threw it back at their opponents. Much more heat than light was generated.

The second side to the triangle is represented by disputes between critical rationalism and scientific realism. There is no way in which this debate can be located in a specific controversy as with the *Positivismusstreit*; to a considerable extent it must be constructed from somewhat more diffuse sources. Realism, as will be seen in more detail later, has long been an available philosophical position with respect to knowledge in general and scientific knowledge in particular; whether it is seen as in opposition to positivism has depended on how broadly or narrowly the latter has been conceived. Popper represents his own philosophy as realist, and sees this as a component of his antipositivism (1976, p. 75). Other writers see no essential incompatibility: Ajdukiewicz remarks that "in the history of philosophy positivism has taken a variety of shades: idealist, realist and neutral" (1973, p. 64). The realism presently at issue has defined itself narrowly and its antipositivism broadly; it is opposed to 'positivist' characterizations of the natural sciences and to their mistaken acceptance by the social sciences. For these purposes it is unnecessary for it to make strict distinctions between critical rationalism and the positivism which the latter rejects. However, to the extent that critical rationalism reacts at all to realism's challenge, the use of the word 'positivism' has not become a shibboleth to anything like the same extent as in the *Positivismusstreit*.

The triangle is completed by the relationship between critical theory and scientific realism. In this case it is hardly appropriate to speak in terms of a debate at all, since there is scarcely any contact between these two traditions; one of the main purposes of the present work is to investigate the implications of establishing such contact. As will be mentioned more fully later, a small number of writers in Britain have begun to discuss aspects of the interrelationships between critical theory and scientific realism in a partial manner: Keat and Urry, from a realist perspective, include a short discussion of critical theory (1975, pp. 218ff.; also Keat, 1981), and Anthony Giddens has produced the only extended essay that compares the critiques of positivism of both critical theory and what he calls the "newer philosophy of science" which includes some discussion of realism (1977, pp. 29–89). German writers within the tradition of critical theory appear to be unaware of, or uninterested in, recent developments in scientific realism; the form in which realism enters controversies to which critical theory is a party is that propounded by critical rationalism. Even within this limited and rather one-sided dialogue, however, accusations of positivism are not entirely absent: Keat and Urry conclude that:

> We regard Habermas's analysis of the empical-analytic sciences as failing to challenge the positivist understanding of scientific knowledge, its view of theories and explanations, and its ontological and epistemological restrictions (1975, p. 227).

Much the same point has been made by Giddens too (1976, p. 68). As far as can be discovered, the compliment has not yet been returned.

If considerable mutual misunderstanding, lack of understanding, and antagonism were generated in the *Positivismusstreit* by divergent use of the term 'positivism', it can be expected that these states may be aggravated when the debate becomes triangular. Nor need this be the stopping-point. Other controversies in which the use of the term 'positivism' has been an integral feature could be brought into the picture. The development of ethnomethodology and phenomenological sociology was characterized by an attack on the rest of conventional 'common-sense' sociology as 'positivistic' (Douglas, 1971; Filmer, 1972); and this charge has been returned not just by writers claiming no affiliation to the movement (Bauman, 1973) but also by its erstwhile adherents (McHugh, 1974). However, although some of the issues surrounding ethnomethodology are strongly related to themes raised by critical theory, especially where the latter draws on the tradition of hermeneutics, no extended consideration can be given in the present context to these controversies. Rather, the aim is to investigate the conception of positivism which is made the object of criticism by each of these three antipositivist tendencies, critical rationalism, critical theory and scientific realism, and to examine the debates between them, in the hope that greater clarity will be gained and that something valuable will emerge from the field of tension formed by their triangular relationship, in particular by giving greater consideration to the least developed side of the triangle, the link between critical theory and scientific realism.

At this point one possible objection may be raised. If the aim is to 'compare and contrast' critical theory and scientific realism, is not this extended and tedious discussion of divergent conceptions of positivism rather a devious way of approaching the subject? The whole debate seems in danger of becoming not just "Hamlet without the Prince", as Giddens has already remarked (1974, p. 18), but Hamlet without the Ghost. Could one not go straight to the heart of the matter and dispense with positivism altogether? This is an attractive suggestion, but one which nevertheless misses the mark. For, as will emerge more clearly later, it is an essential feature of each of these schools of thought that they are antipositivist schools, though this is essential to each of them in a different way. A preliminary remark may be in order at this stage concerning the essential antipositivism of critical theory, which shows that it would be mistaken to treat its relationship to scientific realism in isolation from their opposition to positivism.

Critical theory, as it was developed by Max Horkheimer and his colleagues at the Frankfurt Institute for Social Research in the 1930s, and programmatically summarized in the former's article 'Traditional and critical theory' (1972, first published 1937), saw itself as a continuation of the ideology-critical approach of Marx's critique of political economy. It accepted Marx's critique as essentially valid, and recognized that it gained its political perspective from its addressee, the industrial proletariat, which, however sceptical Horkheimer might be concerning its actual political consciousness and preparedness for struggle, was nonetheless the only potential agent of radical social transformation in the advanced capitalist societies. Critical theory aimed to go beyond the critique of political economy to show how the distorting influence of capitalism reached deeply into all aspects of the culture and forms of thought of bourgeois societies. However, after the move of Horkheimer and other members of the institute to the United States, a shift of ground took place, first clearly apparent in Horkheimer and Adorno's *Dialectic of Enlightenment* (1973), published originally in 1944. They no longer believed that classical Marxism was sufficient to explain the development of advanced capitalist societies or to guide the political orientation of forces capable of transforming them. They also came to believe that the theories of Marx and Engels were partly responsible for the authoritarian and bureaucratic development of the Soviet Union. Unplanned, liberal capitalism had given way to a form of technocratic planning and domination which involved the increasing reification of all aspects of social and cultural life. These tendencies, rather then the capitalist relations of production alone, were now seen as the main obstacle, both in the West and the East, to emancipation and freedom. But the consequence of this for critical theory was far-reaching. As Jürgen Habermas later expressed it, with the repoliticization of the institutional framework of society, "a critical theory of society can no longer be constructed in the exclusive form of a critique of political economy" (1971b, p. 101). Critique had rather to be directed towards the natural scientific and technical forms of thought which on the one hand contributed to the development of the productive forces, allowing a level of productivity never before dreamed of, but on the other hand formed the basis of that technocratic domination which prevented the creation of a truly rational and free society. Critique had to become the 'critique of instrumental reason' or, more generally, the critique of positivism, where 'positivism' comes increasingly to be used to refer to the philosophical expression of the domination of technical rationality. Critical theory is thus not involved 'merely' in an attempt to show that 'positivism' is a mistaken account of the nature of knowledge in the natural and social sciences, but also in an attempt to show that it performs central *ideological* functions in advanced industrial societies. Epistemology is intimately bound up with the theory of society (see Kreckel, 1975, pp. 84ff.; Wellmer, 1971).

It can be seen, then, that critical theory is not incidentally but essentially antipositivist, and hence its relationships with both critical rationalism and scientific realism cannot be analyzed in isolation from the respective forms of antipositivism of each of these three schools of thought. The Ghost may be there only in Hamlet's imagination (but then in that of Marcellus, Bernardo and Horatio too?), but the play is unintelligible without it.

However, the introduction of the theme, central to the concerns of critical theory, of 'positivism as ideology', raises another difficulty for this entire fourth approach to the identification of positivism. It was said above that the presuppositions of this approach were hermeneutic ones, and thus, on Gadamer's interpretation, tied to the practical task of sustaining mutual understanding and self-understanding in the context of the authority of a developing tradition. Criticizing Gadamer, Habermas has argued that there is an inherent conservatism in a standpoint that binds all interpretation to the continuing authority of tradition, and that Gadamer fails to make a distinction between authority and reason (1971a, p. 156). For traditions develop in specific empirical social conditions, and can function as ideological legitimations of power and domination in such social conditions (Habermas, 1970a, pp. 287ff.). Critical reflexion is directed towards the criticism of such traditions as ideology: in such contexts hermeneutic interpretation is transformed into critique of ideology. If we apply this argument to the present problem of the identification of positivism, we find that a purely hermeneutic approach to the development of antipositivist traditions will be unable to do justice to the critique of positivism advanced by critical theory, precisely because the latter attempts to include critical rationalism as subject to its critique. Mutual understanding seems to be ruled out in advance by such a stance (Kreckel, 1975, p. 97), which would in fact constitute a fifth, ideology-critical, approach to the identification of positivism. However, a purely hermeneutic approach, excluding questions of ideology, might be thought to give too much away to critical rationalism at the outset.

There is no way in which this difficulty can be either by-passed or decided in advance, and eventually it will have to be met on its own terms. The plan of campaign is as follows: in the following two parts, the three antipositivist traditions and their relationships to each other will be reconstructed, firstly with respect to the theory of knowledge in the natural sciences, secondly with respect to the theory of knowledge in the social sciences. In the first of these two sections, there will be more discussion of the epistemology of the natural sciences than might be thought, by either adherents or opponents of the unity of science, to be appropriate in a work about sociology; but however positivism has been interpreted, the question of the relationships between the natural and the social sciences has always been a central one, and, as Giedymin puts it:

> one's methodological diagnosis of social science in relation to natural science depends, of course, as much on how one sees the latter as on how one sees the former (1975, p. 282).

While it will be impossible to avoid problems associated with the concept of ideology in either of these two sections, a final part will attempt to draw the threads together by focussing specifically on the relationship between a hermeneutic approach to philosophical traditions and the critique of ideology.

PART II

ANTIPOSITIVISM IN THE PHILOSOPHY OF THE NATURAL SCIENCES

CHAPTER 2

THE ANTIPOSITIVISM OF CRITICAL RATIONALISM

In an article written as a reaction to the publication of *Der Positivismusstreit in der deutschen Soziologie*, Karl Popper rejects the application of the term 'positivism' to his own thought.

> Words do not matter, and I do not really mind if even a thoroughly misleading and mistaken label is applied to me. But the fact is that throughout my life I have combatted positivist epistemology, under the name 'positivism' . . . I have fought against the aping of the natural sciences by the social sciences, and I have fought for the doctrine that positivist epistemology is inadequate even in its analysis of the natural sciences which, in fact, are not 'careful generalizations from observation', as it is usually believed, but are essentially speculative and daring; moreover, I have taught, for more than thirty-eight years, that all observations are theory-impregnated, and that their main function is to check and refute, rather than to prove, our theories. Finally I have not only stressed the meaningfulness of metaphysical assertions and the fact that I am myself a metaphysical realist, but I have also analysed the important historical role played by metaphysics in the formation of scientific theories (Adorno *et al.*, 1976, pp. 298ff.).

Popper identifies 'positivism' with the logical positivism of the Vienna Circle, and to understand critical rationalism's particular form of antipositivism, it is necessary, as Popper himself suggests, to go back to his opposition to logical positivism in his first major work, *Logik der Forschung* (Popper, 1934, 1968).

The members of the Vienna Circle thought of their movement as an attack on metaphysics. There had been attacks on metaphysics before in the history of philosophy, notably that summed up in the famous final paragraph of David Hume's *Enquiry Concerning Human Understanding* (1902, p. 164), and that of Comte, which constitutes the one link between 'the Positive Philosophy' and Logical Positivism which no account of positivism would deny, even if it would deny any other. It might be thought that positivism's attack on metaphysics was more radical than that of other philosophical schools; thus Ayer states that:

> An explicit rejection of metaphysics, as distinct from a mere abstention from metaphysical utterances, is characteristic of the type of empiricism known as positivism (1962, p. 135).

And Viktor Kraft reports that "metaphysics was to be completely eliminated" (1953, p. 15). But what was really different about logical positivism was the grounds upon which the attack on metaphysics was based. As Ayer puts it:

> The originality of the logical positivists lay in their making the impossibility of metaphysics depend not upon the nature of what could be known but upon the nature of what could be said (1959, p. 11).

The criteria for what could meaningfully be said were derived by transforming Russell's logical atomism into the Principle of Verifiability.

The central assumption of logical atomism was that all statements, however complicated, could be analyzed as combinations of absolutely simple, elementary statements. These elementary statements corresponded to the atomic facts which made up the world. The truth of any compound or molecular statement depended on the truth of the elementary statements into which it could be analyzed, making use of the logical connectives, conjunction, disjunction, implication and negation, with which elementary statements could be connected together. The truth distributions of molecular statements could be summarized in truth tables. On this basis, different categories of molecular statements could be distinguished:

(a) There were, firstly, those molecular statements which agreed either with every truth distribution of its constituent atomic statements or with no distribution. The latter were identified with contradictions and the former with tautologies. It was held that all the statements of logic and mathematics were of these types.

(b) There were, secondly, those molecular statements which agreed, or were capable of agreeing, with some truth distributions of its constituent atomic statements but not with others. The truth of a molecular statement, in other words, depended on the truth of the atomic statements into which it could be analyzed. These statements were identified with empirical statements, statements about the atomic facts of the world and their combinations.

(c) There were, finally, statements which were neither tautologies or contradictions, nor statements whose truth could be determined by reference to the truth of the atomic statements of which they were composed. It would be more accurate to say that such statements were not composed of elementary statements at all, that there was no way in which they could be analyzed. There was thus no way in which their truth could be determined. These statements were identified with metaphysics (Ayer, 1959, pp. 11f.).

Logical atomism provided a logical framework for the attack on metaphysics, but did not by itself provide a criterion for identifying metaphysical statements. This was provided by grafting on to logical atomism a version of sensationalism that could be derived from Hume or Mach. The atomic statements into which molecular statements were analyzed were given no very definite interpretation by Wittgenstein in the *Tractatus* (Pears, 1971, p. 71). The members of the Vienna Circle, however, took it for granted that these elementary statements would be the direct reports of observation, 'protocol statements' as they came to be called. This combination of logical atomism and sensationalism generated the 'principle of verifiability', which was expressed as a criterion of meaningfulness; in its simplest form, the principle states that the meaning of a proposition is its method of verification. Verifiability was taken to be a standard for according to or withholding from a system of propositions the stamp of having sound foundations. Logic and mathematics were meaningful, although they tell us nothing but what was implicit in what we knew already, because their procedures defined a way of verifying any statement made within them. Empirical science was meaningful because it was based on a foundation of sensory observation, which provided a method of verification for any empirical statement. But metaphysics could be rejected as meaningless, because there was no way in which its statements could be verified. In the case of paradigm metaphysical statements like "Reality is the Absolute", there was no possible experience that could have a bearing on its truth, let alone verify it (Passmore, 1968, p. 368); or, as the pragmatists had it, "there was no difference that *made* a difference between asserting it and denying it" (Achinstein and Barker, 1969, p. 5).

Without going further into the detail of logical positivist ideas, it is possible to distinguish two different kinds of problem that can emerge in relation to them. The first are problems which arise within logical positivism as it tries to develop its programme. Such problems have included the attempt to maintain the doctrine that mathematics can be derived from logic, against the criticism of formalism and intuitionism; the attempt to maintain the empiricist distinction between analytic and synthetic propositions, against the criticism of such writers as Quine (1961); and the attempt to come to some conclusion on the vexing question of the status of protocol statements (Passmore, 1968, pp. 372ff.). These are the problems which the members of the Vienna Circle were essentially concerned with. There is a second kind of question, however, that concerned the status of logical positivism's own statements. The most obvious form in which this problem arose was in the question of the status of the principle of verifiability itself, which appeared, on the face of it, to be neither a tautology nor an empirical statement. If it was neither of these, then strictly speaking it was meaningless, and should be rejected as metaphysical nonsense. And what applied to the principle of verifiability could apply to any other philosophical statements made by the logical positivists.

There were at least four ways of coping with this difficulty. The first was to assimilate philosophy to logic, more specifically to the logic of science. There could be, on this formulation, no other task for philosophy than an analysis of the logical structure of scientific theories, of the principles of induction used by scientists, of the logic of explanation, and so forth. Any attempt to go beyond this would result in meaningless metaphysics. The second way was to assimilate philosophy to empirical science. What were once thought to be philosophical problems would then turn out to be problems concerning the empirical psychology of cognition, the empirical sociology and psychology of scientific behaviour, or, in many cases, empirical natural science. This approach is encapsulated in the term 'scientific philosophy' (Reichenbach, 1951). The third way was to accept that the statements of philosophy were indeed metaphysical, and hence nonsense. Thus Wittgenstein, towards the end of the *Tractatus*, took this most consistent, and (only apparently) most self-destructive step, saying:

> My propositions are elucidatory in this way; he who understands me finally recognises them as senseless, when he has climbed out through them, on them, over them. (He must so to speak throw away the ladder, after he has climbed up on it.) He must surmount these propositions; then he sees the world rightly (1961, 6.54).

The fourth way, which can be taken as a way of avoiding the embarrassment of the third, was also taken by Wittgenstein, as well as by others. It was to argue that philosophy was not a body of statements at all, and hence not subject to the meaningfulness test of the verifiability principle. Philosophy could only be preserved if it was conceived of as an activity, an activity which Wittgenstein outlined as follows:

The right method of philosophy would be this: to say nothing except what can be said, i.e. the propositions of natural science, i.e. something that has nothing to do with philosophy: and then, always, when someone else wishes to say something metaphysical, to demonstrate to him that he had given no meaning to certain signs in his propositions (1961, 6.53; cf. 4.112).

Philosophy is the activity of making clear to people what they can, and what they cannot, say.

We are now in a position to turn to Popper's *Logik der Forschung*. As Popper explains in his autobiography (1976, pp. 78ff.), this was written as a development of his Ph.D. thesis while he was in close contact with the Vienna Circle, at first through their writings and later through personal acquaintance, although he was never invited to Schlick's private seminar which Popper thought of as the core of the Circle, and hence disclaims ever having been a member of it. The book, despite its publication in the series *Schriften zur wissenschaftlichen Weltauffassung*, which was edited by Schlick and Frank and mainly included books written by members of the Vienna Circle, was indeed a sustained critique of many of the central tenets of logical positivism. It is often suggested that Popper merely substituted falsifiability for verifiability, thus 'solving' the problem of induction, but otherwise left everything as it was; and indeed this was how some members of the Vienna Circle received his book. In fact, however, as Popper himself stresses, *Logik der Forschung* is far more than an alternative account of the logic of scientific method; it contests the whole programme of logical positivism, the attempt to distinguish science from metaphysics in order to reject the latter as meaningless. He wrote:

Positivists . . . believe they have to discover a difference, existing in the nature of things, as it were, between empirical science on the one hand and metaphysics on the other. They are continually trying to prove that metaphysics by its very nature is nothing but nonsensical twaddle – 'sophistry and illusion', as Hume says, which we should 'commit to the flames' (1968, p. 35).

Popper rejects this attack on metaphysics:

In contrast to these anti-metaphysical stratagems – anti-metaphysical in intention, that is – my business, as I see it, is not to bring about the overthrow of metaphysics. It is, rather, to formulate a suitable characterization of empirical science, or to define the concepts 'empirical science' and 'metaphysics' in such a way that we shall be able to say of a given system of statements whether or not its closer study is the concern of empirical science (1968, p. 37).

In other words, Popper aimed to demarcate, not science from metaphysics, but science from non-science, and there was no intention to show that non-science was meaningless.

Popper had a number of substantive reasons for rejecting the logical positivists' attack on metaphysics. Firstly, he did not think metaphysics was meaningless, and did not think it made much sense to say that it was. He argued that verifiability could not be a criterion of meaningfulness, since one had to be able to understand a statement in order to know whether it was verifiable, and if one could understand it, it must be meaningful. This was a common objection to the principle of verifiability and, although it was later acknowledged as a problem by logical positivists such as Ayer (1962, p. 6), it did rather miss the point of the logical positivist programme, which depended on trying to show that those who uttered metaphysical statements did not really understand those statements, even if they thought they did, since no sense could be attached to them. A second substantive objection was contained in Popper's rejection of logical atomism, since he did not believe that all propositions could be analyzed as a compound of elementary statements. In particular he did not believe that scientific theories or natural laws were so analyzable, and he rejected the inductive principles on which a logical atomist account of scientific laws had to be based. An attack on metaphysics based on the principles of logical atomism would be forced to reject natural science as metaphysics as well (1968, p. 36). Even if Popper's rejection of induction and introduction of the deductivist principle of falsifiability could not be accepted by a consistent logical positivist, this problem of the justification of induction was one which they, too, had to face (Halfpenny, 1976, pp. 133ff.). A third objection was to sensationalism, which was taken by Popper to be the claim that empirical science rested on the ultimate incorrigibility of reports of direct sensory experience, the positivists' 'protocol sentences'. Popper rejected the notion that there were any 'ultimate' statements (1968, p. 47), and substituted the idea of 'basic statements' which always remained revisable by inter-subjective testing. This was another arrow in his assault on the principle of verifiability, and hence on the positivist attack on metaphysics. Again, this problem was recognized by the Vienna Circle too, and the '*Protokollsätze* debate' was concerned precisely with the incorrigibility or otherwise of protocol sentences, and with the question of what the protocol sentences were protocols *of* (see Passmore, 1968, pp. 372ff.; Halfpenny, 1976, pp. 57ff.).

However, underlying these substantive objections which Popper had to logical positivism was a more 'deep-seated difference' between his position and that of the positivists concerning the nature of philosophy and its relationship to science. He wrote:

> The positivist dislikes the idea that there should be meaningful problems outside the field of 'positive' empirical science – problems to be dealt with by a genuine philosophical theory. He dislikes the idea that there should be a genuine theory of knowledge, an epistemology or a methodology (1968, p. 51).

For a logical positivist, as we have seen, such problems are either "pseudo-problems" (Carnap, 1967), which gave rise to meaningless metaphysics, or else were to be treated as branches either of empirical science or of logic.

Popper rejected the idea that problems of epistemology or methodology were pseudo-problems, just as he rejected the similar disparagement of metaphysics in general, although he did not expect the positivists to be convinced (1968, p. 52). His argument against the second possibility was particularly significant. If epistemology was not to restrict itself to logical analysis, it would have to be a branch of empirical science.

> This view, according to which methodology is an empirical science in its turn – a study of the actual behaviour of scientists, or of the actual procedure of 'science' – may be described as 'naturalistic'. A naturalistic methodology (sometimes called an 'inductive theory of science') has its value, no doubt. A student of the logic of science may well take an interest in it, and learn from it. But what I call 'methodology' should not be taken for an empirical science. I do believe that it is possible to decide, by using the methods of an empirical science, such controversial questions as whether science actually uses a principle of induction or not (1968, p. 52).

In other words, the fact that scientists generally *do* follow certain procedures cannot itself be a rational justification of those procedures. Methodology cannot be purely descriptive, it must raise claims of a normative kind.[3]

Popper's own conception of methodology is that "methodological rules are here regarded as conventions" (1968, p. 53).

I shall try to establish the rules, or if you will the norms, by which the scientist is guided when he is engaged in research or in discovery (1968, p. 50).

These rules do thus have a descriptive status: they describe the 'game of science'. But it follows from Popper's critique of naturalistic methodology that they cannot *just* have a descriptive status; they are not intended merely to describe one possible game of science among others (Wellmer, 1967, p. 90), but to provide a normative methodology for *the* concept of empirical science. They are thus conceived of as conventions, which if accepted and followed will characterize what is legitimately called science.

These methodological rules, or conventions, are hierarchically arranged. The highest level rule is the principle of falsifiability, which acts as the criterion of demarcation of science from non-science. It is Popper's "proposal for an agreement or convention" which will be a "suitable characterization of empirical science" (1968, p. 37). Then there is the rule that lays down that no scientific statement is to be protected from falsification by "conventionalist stratagems" (1968, p. 82), and this is followed by a series of other rules concerning degrees of falsifiability, the acceptance of basic statements, and so on (see Johansson, 1975). It is not necessary to describe the system of methodological rules here. The more important question from the present point of view is how such rules, such proposals for conventions, are to be justified, for their justification cannot come from the fact that they are followed by scientists; it is in fact compatible with Popper's account that scientists do not always follow them.

Popper's answer to this depends on the concept of 'purpose':

As to the suitability of any such convention opinions may differ; and a reasonable discussion of these questions is only possible between parties having some purpose in common (1968, p. 37).

The justification for conventions over methodological rules will depend on the purpose that is set for science. Popper gives a brief characterization of the purpose he sees in science. He rejects the view that the purpose of science can be to provide a system of absolutely certain, irrevocably true statements, and stresses instead "the characteristic ability of science to advance" (1968, p. 49). He hopes that his proposals

may be acceptable to those who value not only logical rigour but also freedom from dogmatism; who seek practical applicability, but are even more attracted by the adventure of science, and by discoveries which again and again confront us with new and unexpected questions, challenging us to try out new and hitherto undreamed-of answers (1968, p. 38).

Finally, it is significant that he does not try to justify this statement of the aims of science, or to represent them as the true or essential aims of science. Rather he characterizes his statement of them as a personal choice of values:

The choice of that purpose must, of course, be ultimately a matter of decision, going beyond rational argument (1968, p. 37).

The opposite course, to talk of "essential or true" aims of science, he describes as "a relapse into positivist dogmatism" (1968, p. 38).

We may summarize Popper's critique of positivism as follows: behind the differences of substance which separate him from the Vienna Circle, most notably his dismissal of induction and his substitution of a falsificationist methodology, there lies what is intended to be a fundamental difference of conception of philosophy and its relationship to science, according to which Popper claims to rescue a true philosophy from the destructive positivist attack on metaphysics. This true philosophy is to take the form of casting epistemological problems as problems of a theory of scientific method, which is to be at the same time a theory of scientific progress; and the guideline for this theory is to be found in the idea that scientific progress is made possible by the adoption of a system of methodological rules as conventions. These conventions, and in particular the most fundamental one that all statements of empirical science should be falsifiable-in-principle, will also serve as a criterion of demarcation of science from non-science, which is to take the place of the rejected positivist criterion of demarcation of sense from non-sense, namely the principle of verifiability.[4]

CHAPTER 3

THE ANTIPOSITIVISM OF CRITICAL THEORY

In this chapter, three aspects of the antipositivism of critical theory will be addressed. In the first place, it is necessary to examine the way in which the concept of positivism is widened so that philosophical tendencies, which according to their own self-understanding are antipositivistic, and in particular Popper's critical rationalism, are brought within its scope. Secondly, critical theory's own critique of this widened notion of positivism will be outlined. Thirdly, an idea will be given of what critical theory itself intends to put in the place of positivism, as it relates to the theory of the natural sciences.

3.1. Critical Theory's Identification of Positivism

Every commentator on positivism, as indeed on any other school of thought, would be hard put to avoid recognizing a certain amount of internal diversity. Thus even Simon, who as was seen above wanted to restrict the application of the term positivism to "the doctrine founded by Auguste Comte", had to recognize "its amenability to different emphases and interpretations" (1972, p. 11), and spends a considerable proportion of his book documenting divergences and schisms between various groups of positivists. Admittedly, the internal diversity which Simon deemed admissible had its limits set by its adherents' confession of positivist faith, but even this is not a criterion which operates itself automatically, since not every such confession of faith can be equally worthy of acknowledgement. Where self-identification is no longer a possible criterion, the question of a restriction on internal diversity is even more problematic.

It is then possible to generate elaborate classifications of different types of positivism, in an effort to come to grips with the problem of internal diversity. One such elaborate scheme is worked out by Halfpenny; he initially demarcates positivism by the central tenets of empiricism and a commitment to the unity of science, and after investigating the multifariously different forms which both of these tenets can take, in particular empiricism, which he conceives as a "multidimensional epistemological space", concludes that positivism is a "many-faceted doctrine, containing at least as many variants as produced by the intersection of the different Unity of Science$_p$ theses with the different versions of empiricism" (1976, pp. 64f.).

One feature of such attempts to deal with the internal diversity of positivism by means of a classification of different varieties is that they are ahistorical, or static. They find it difficult to do justice to the fact that positivism, like other schools of thought, has a history, and that this history is not to be understood as the succession of different variants, but as a dynamic development through argument, in the course of which positivist thought changes its shape. This historical development can be more fully grasped if one penetrates beneath the surface succession of variants to the motor of this argumentative dynamic, the problematic, to adopt the contemporary vogue, which drives positivism on. This is the approach taken by Herbert Schnädelbach, a writer much influenced by Adorno and Habermas, in a book which, more systematically than the polemical contributions to the *Positivismusstreit*, examines the development of positivism and the relationship of critical rationalism to that development (Schnädelbach, 1971). It will be instructive to follow the main argument of Schnädelbach's work, which can stand as one of the most thorough justifications of critical theory's widened notion of positivism.

In commentary on positivism there is some discord over whether positivism is essentially a sceptical, destructive doctrine or whether it is essentially an affirmative one. The former interpretation rests its case on the attack on metaphysics: what can be known or what can be said is limited by highly restrictive criteria, resulting from a destructively sceptical attitude to what, it is suspected, cannot be known or said. The latter interpretation is derived, on the other hand, from an emphasis on 'the positive' contained in positivism's name and from which it was etymologically derived: the intention is affirmatively to ground knowledge on a positive basis, which, whatever else has to be rejected, will stand as a secure foundation for scientific knowledge. The central opposition in this second interpretation is that between the negative and the positive. Thus Marcuse writes:

> Positive philosophy was a conscious reaction against the critical and destructive tendencies of French and German rationalism, a reaction that was particularly bitter in Germany. Because of its critical tendencies, the Hegelian system was designated as 'negative philosophy' (1955, p. 325).

Schnädelbach's solution to this apparent discord is to argue that positivism is in fact based on a pattern of thought which unites both of these opposing motifs: the sceptical one and the affirmative one (1971, p. 8). This combination of opposites gives positivism its dynamic, for, although the affirmative motive is dominant over the sceptical one, the intention to provide a firm foundation for knowledge dominant over the intention to reject metaphysics, the recurring pattern throughout the history of positivism is that the scepticism is directed against whatever is proposed as a positive foundation at any given moment, and the dynamic development which this generates proceeds until positivism itself dissolves.

Comte distinguished five meanings of the word 'positif', which were to characterize the constitutive features of the 'philosophie positive' (1875, Vol. 1, p. 45). These were: 'le réel', 'l'utile', 'la certitude', 'le précis', and 'le contraire de négatif'. The most basic of these meanings was the first, the real in the sense of the factual, and the dependence of the other four meanings on the first is symptomatic of the epistemological principle of positivism: that useful, certain, precise, and constructively valuable knowledge can only be founded on the basis of the factually real, the world of facts. But this principle does not sufficiently distinguish positivism from other epistemological tendencies, and Schnädelbach sets out what he sees as the programme, the systematic premisses and methods which historically conditioned the structure of positivist thought (1971, pp. 13ff.).

As an epistemological position, positivism stands in the tradition of empiricist attempts to provide a new foundation for knowledge on the basis of that which cannot be doubted. This attempt follows the cartesian programme, and what qualifies it as empiricist is the idea that the indubitable basis must be presented in the realm of sensory experience. The idea that there must be something that is absolutely certain, otherwise nothing is certain, is itself a presupposition. The positive basis acts as the limit of scepticism, both by showing that a universal scepticism in relation to all human knowledge is unjustified, and by providing a basis against which any sceptical objection to a specific item of knowledge can be tested; and any claim to knowledge that turns out not to be decisively testable against that basis is identified with metaphysics and excluded from human knowledge.

> The restriction of scepticism, the exclusion of idle speculation, and the guarantee of intersubjectivity are the epistemological goals which positivism shares with the empiricist tradition, and it remains in that tradition to the extent that it follows the empiricist principle only to allow those ideas to be considered as valid knowledge which are founded in experience (1971, pp. 14f.).

This principle necessarily had to apply, not just to any knowledge whose status as knowledge was in need of justification, but also to the method of epistemology itself: for, just as much later logical positivism had to face the problem of its own cognitive status according to its own criteria, as was seen above, so empiricism at a very early stage had to recognize that its own claim to be a true account of knowledge also was subject to empiricist criteria of adequacy. This resulted in the empiricist programme, often criticized as circular, which aimed to find an indubitable experiential basis for knowledge by means of empirical analysis of the process of cognition.

Schnädelbach describes this programme as an empiricist version of the cartesian attempt to provide a grounding for knowledge in the medium of reflection, i.e. the turning back of the knowing consciousness on to itself; but whereas in the cartesian tradition this reflection follows the procedure of logical analysis, in the empiricist tradition from John Locke it undergoes an empiricist re-interpretation and proceeds by an empirical analysis of consciousness, 'the historical, plain method' as Locke terms it (1959, Vol. 1, p. 27). Reflection becomes a form of inner perception, epistemology becomes psychologistic, and the typical difficulty for this epistemology of relating the 'ideas', 'impressions', or 'sensations' to an objectivity in itself, transcendent to consciousness, is highlighted by the move in Berkeley's philosophy through subjective idealism to a spiritualistic or theological metaphysics, to which the empiricist reaction is to confine itself entirely to what is immanent to consciousness, either in some form of sensationalism, or in some form of linguistic phenomenalism, both of which are attempts to characterize the *basis* on which knowledge can be built.

However, the nature of this basis is also dependent on another presupposition which empiricism makes, namely that concerning the nature of experience itself. In order to avoid the circularity or infinite regress involved in the attempt to erect an empiricist epistemology on an empirical analysis of cognition, this concept of experience has to be introduced a priori. In fact,

Locke's partiality for the 'historical, plain method' already implies the conviction that the simple description of objective processes, abjuring all formation of hypotheses, is a legitimate and necessary procedure (1971, p. 19).

Consequently experience is conceived of as a receptive process: consciousness is not in a position itself to create 'ideas' or 'impressions' as the material of knowledge, but merely receives them from without. This model of receptivity generates in addition 'cathartic' motives, the demand that consciousness should purify itself of all distorting opinions or attitudes in order that unadulterated truth be able to gain access. This presupposition is the basis of Bacon's theory of the idols. Further, the model of receptivity is filled out by a theory of the senses, either, as in Locke, that sense-perception is a direct physical process of the influence of things on the senses, or, when the thesis of immanence casts doubt on this direct realism, in a form which lends to the evidential certainty (*Evidenz*) of the senses the maximum degree of truth and reliability.

It is assumed that in the process of reception the information which is available to be received by the senses is liable to be distorted by natural, individual, or cultural factors. There is in empiricism a systematic mistrust of all elements of knowledge that can be traced to the subjective activity of the cognitive process, a mistrust which, according to Schnädelbach, is a necessary consequence of the presupposed objectivist concept of truth.

The possibility of the critical reductive procedure following criteria of the empiricist model of receptivity is dependent in turn on whether it is possible clearly to distinguish between received and produced components of knowledge. This is the systematic location of the leading idea which Quine identifies as one of the 'dogmas of empiricism': that of the possibility of differentiating between those judgements which elucidate knowledge and those which extend it, between analytic and synthetic statements, which in empiricism follows precisely the difference between productivity and receptivity. Only that which is received is capable of extending knowledge synthetically (1971, pp. 21f.).

The presuppositions of the receptivity-model provide empiricism and positivism with a central condition of their immanent further development, since they demand a search for that which is received without any admixture of subjective productivity, that which Avenarius termed 'pure experience' (1888–1890), in order that it may fulfil the role of the basis of knowledge. The history of empiricism, Schnädelbach suggests, may be written as the series of attempts to achieve an increasingly pure distillation of received experience from the mass of available conceptions, and

> the positivist theories of the basis are the result of a process of the step-by-step separating out of the posited from the given, in which whatever is put forward as supposedly purely given provokes, again and again, a further process of separation (1971, p. 23).

The counterpart in the receptivity-model to the purely given or received is the idea of an empty consciousness, for example in the form of Locke's *tabula rasa*, an idea generated in his case by means of a critique of the doctrine of innate ideas. Together, these two sides of the model of receptivity constitute two basic ideas of the empiricist epistemology, and they also involve a naturalistic interpretation of the cognitive process, which assumes that cognition obeys its own laws, natural laws of psychology which like any other natural laws may be discovered by empirical investigation.

The final presupposition which structures the 'problematic' of positivism is a thoroughgoing nominalism.

> The empiricist principle itself follows from the metaphysical thesis that only particulars are real, and that it is logically impossible to specify a purely conceptual principium individuationis. If therefore the particular facts can neither be completely grasped in a concept nor be conceptually-deductively anticipated, then no purely mental operation can yield primary access to reality, but only sense-experience (1971, p. 28).

This nominalism generates the reductionist direction of empiricist epistemology and thus leads to Quine's second 'dogma of empiricism': the conviction that only indivisible simples, whether they be atomic sense-experiences or elementary statements, can constitute the basis of possible reconstructions of valid knowledge.

To complete his account of the structural elements of positivist thought, Schnädelbach mentions the traditional conflict to which the presuppositions of empiricism give rise: that between the thesis of immanence and the model of receptivity.

> The empiricist-reflexive procedure of providing grounds for knowledge qualifies the basis-realm from the outset as the set of determinate mental experiences. If these experiences are understood, on the model of receptivity, as produced by causes transcendent to consciousness, the problem arises of how this notion of a cause transcendent to consciousness can be legitimated from the standpoint of a theory that has committed itself epistemologically to the thesis of immanence (1971, p. 29).

This is a continual and insoluble problem within the presuppositions of empiricism, for the idea of immanence can only be defined in opposition to that of transcendence, and the idea of that which is transcendent to consciousness must be at least meaningful if the immanence-thesis is to have any sense. This conflict has historically given rise to such widely divergent philosophies as the neutrality-thesis of Mach and Avenarius and the absolute idealism of Fichte.

This account of the presuppositions of empiricism and the philosophical problems which they generate is intended by Schnädelbach not as a static definition of a philosophical school but rather as a dynamic identification of a philosophical tendency in terms of the motor of its internal development. Since there can be no pretension here to a detailed history of either empiricism or positivism, it is unnecessary to follow Schnädelbach through his examination of the working out of this internal development in, firstly, the empirio-criticism of Ernst Mach, and, secondly, the logical positivism of Schlick and Carnap. What is essential is to grasp the overall conclusion of these detailed analyses. This is that in each case the attempt, forced on positivism by its empiricist presuppositions, to separate out a pure realm of the given or the received which is to be free from the subjective distortion of positing or productivity, has to be abandoned. The scepticism of positivism eventually overcomes the affirmative intention of discovering an indubitable basis on which knowledge could be grounded, and becomes directed not only against every proposal that is advanced for such a basis but also at the positivist programme itself. Positivism dissolves, and what it dissolves into is conventionalism (1971, pp. 9, 131ff.).

3.1.1. Relationships Between Positivism and Conventionalism

Since the analysis of conventionalism plays a vital role in critical theory's identification of positivism, a brief discussion of relationships between these two notions, in the context of the philosophy of the natural sciences, is here in order. Different writers on positivism treat this relationship very differently. Conventionalist writers themselves, for example Duhem, Poincaré, and Le Roy, represented themselves as antipositivist or post-positivist. To the extent that conventionalism is put forward as a separate philosophical position at all in surveys of thought in the philosophy of the natural sciences, their claim is not unanimously accepted. Losee gives conventionalism a chapter of its own in his historical introduction to the philosophy of science, but associates it with 'mathematical positivism' (1972, Chapter 11). Keat and Urry distinguish sharply between positivism and conventionalism, and, as will be more fully discussed later, associate their scientific realism more closely in its aims with positivism than with conventionalism (1975, Part One). Kołakowski incorporates a chapter on conventionalism into his book on positivist philosophy, and his argument, although brief, comes closest to that of Schnädelbach, for Kołakowski recognizes the complexity of the relationship:

> Conventionalism represents an extension of positivist philosophy, but in one sense it is also a refutation of it, the expression of a self-destructive tendency inherent in it (1972, p. 176).

To understand this, a brief account of conventionalism drawn from Kołakowski is illuminating:

> The fundamental idea of conventionalism may be stated as follows: certain scientific propositions, erroneously taken for descriptions of the world based on the recording and generalization of experiments, are in fact artificial creations, and we regard them as true not because we are compelled to do so for empirical reasons, but because they are convenient, useful, or even because they have aesthetic appeal. Conventionalists agree with empiricists on the origin of knowledge, but reject empiricism as a norm that allows us to justify all accepted judgements by appealing to experience, conceived of as a sufficient criterion of their truth. Or, to put the same point somewhat more accurately, the data of experience always leave scope for more than one explanatory hypothesis, and which one is chosen cannot be determined by experience. Rival hypotheses accounting for a given aggregate of facts may be equally sound from a logical point of view, and hence our actual choices are accounted for by non-empirical considerations (1972, p. 158).

Although, as Keat and Urry (1975, p. 60) point out, there are a number of different strands of thought in this characterization, there is one clear idea in it which relates it to Schnädelbach's account of positivism, namely that conventionalism appears to reject positivism's programme of making a sharp divide between the received and the produced components of scientific knowledge, between the given and the posited, in order that scientific knowledge may be securely built upon the former by cathartically excluding the latter. As Kołakowski too points out, "what conventionalism sets out to criticize was this notion of 'the given'" (1972, p. 177). From the history of positivism's attempts to distil out pure experience, conventionalists claim to learn that there is instead something irredeemably 'produced' about all human knowledge, something that necessarily comes from the side of the human knowing subject, and it captures this 'something' in the concept of the conventional.

However, as the present focus of interest is critical theory's extension of the concept of positivism to include critical rationalism, there is one confusing aspect of this notion of the conventional which must be mentioned before the argument can proceed. For despite Popper's use, in *The Logic of Scientific Discovery*, of the word 'convention' to describe the status of his proposed methodological rules, he explicitly rejects the appellation 'conventionalist' for his own philosophy of the natural sciences. His argument is as follows: conventionalists such as Duhem had seen, quite correctly, that in a scientific experiment it was not the case that a single scientific proposition was compared with the results of empirical observation, but rather a system of propositions. Far from it being the case that a hypothetical law was being tested against experimental results, those laws themselves were presupposed in experimentation and were required to determine what an observation, or more specifically, a scientific measurement is. Thus it could never be necessary to reject a law on the basis of experience, since unexpected experimental results could always be accommodated by the adjustment of other parts of the system of propositions which formed the theoretical background to the experiment. The criteria for the acceptance of a system of laws could not, therefore, be purely empirical, and the conventionalist argues, according to Popper, that only one other principle

> can help us to select a system as the chosen one from among all the other possible systems: it is the principle of selecting the simplest system – the simplest system of implicit definitions; which of course means in practice the 'classical' system of the day (1968, pp. 80f.).

Since any system may be 'saved' from falsification by the use of 'conventionalist stratagems', conventionalism has, Popper believes, conservative implications for scientific development, and it is on these grounds that he rejects conventionalism, rather than on grounds of its logical inadequacy; in fact, he regarded it as 'a system which is self-contained and defensible' (1968, p. 80). But if this is so, Popper's only way of rejecting conventionalism, as he sees it, is by taking a *decision*; the decision not to apply its methods. We decide that if our system is threatened we will never save it by any kind of *conventionalist* stratagem (1968, p. 82).

In other words, conventionalism is avoided by the decision to adopt, as a convention, the methodological rule that systems of scientific statements should not be protected from falsification by conventionalist stratagems.

When Popper's rejection of conventionalism is put in this way, its oddness is immediately apparent. The situation may be clarified by a distinction between two levels of conventionalism, both of which are implicit in the classical conventionalism of Duhem and Poincaré and also in Popper's critical rationalism, but which are given different emphases in the two systems of ideas. One might label the distinction one between '*thesis-conventionalism*' and '*methodological conventionalism*'.[5] Thesis-conventionalism argues that there is something necessarily conventional in the system of statements which make up scientific knowledge or which are put forward as potentially scientific knowledge: experience alone cannot be the criterion for rejection or acceptance of such systems. Methodological conventionalism argues that the additional, humanly produced criteria for the acceptance or rejection of systems of statements are themselves conventional. These two aspects of conventionalism need not necessarily be combined into a single doctrine, but in practice it is not easy to disentangle them. Thus Popper's theory of scientific method quite clearly qualifies as an example of methodological conventionalism, but there is also more than a whiff of thesis-conventionalism in his theory of basic statements, according to which statements are accepted as basic, and hence capable of being used as falsifying instances in the testing of scientific hypotheses, by the consensual decision of members of the scientific community to withdraw them from further test: to that extent they too are conventions (1968, pp. 104ff.). Whether a more classically-oriented thesis-conventionalist is also a methodological conventionalist will depend on the nature of the arguments for the principle or principles, such as simplicity, on the basis of which systems of scientific statements are to be accepted or rejected.

This distinction is not put forward in the first instance with the intention of solving any philosophical problems but with the intention of clarifying the legitimacy of the use of the label 'conventionalism' in the context of a discussion of critical rationalism. There is a certain degree of formalism in this argument, but also a degree of superannuation, given that later critical rationalists, such as Imre Lakatos, have, unlike Popper, not shunned the identification of critical rationalism as a form of conventionalism (Lakatos and Musgrave, 1970, p. 104). The clarification, if it is one, may however be worthwhile, in the light of Popper's own continued use of the term conventionalism only in the context of his rejection of what he refers to as 'conventionalist stratagems', or, in Hans Albert's phrase, "strategies of immunization" (Popper, 1972, p. 30). This restricted usage makes it possible to forget the centrality of the concept of 'convention' to describe the status of methodological rules in the original formulations of *The Logic of Scientific Discovery*, a possibility to which Lakatos' discussion is a salutary correction. In what follows, the term 'conventionalism' will be used to refer to both thesis-conventionalism and methodological conventionalism to the extent that the arguments outlined are equally applicable to both.

Schnädelbach's discussion of the relationship of conventionalism to positivism can be divided into two themes: the location of the concept of 'convention' in the structure of positivism's problematic; and the cognitive status of 'conventions' themselves. In the first place, the concept of 'convention' was introduced, as has been noted above, to refer to those elements of knowledge which are humanly produced or 'subjective', once it was realized that the positivist programme of eliminating such elements from scientific knowledge could not be carried out. However, Schnädelbach argues, this conceptual innovation should not be interpreted as an abandonment of the positivist problematic, but rather the result of positivism's increasing, but still partial, self-awareness. On this interpretation,

> conventionalism would then be the form of positivism's self-consciousness, which itself results from the process of the latter's development and immanent dissolution (1971, p. 9).

More specifically, the model of providing grounding or justification for knowledge which positivism derives from its empiricist presuppositions is maintained intact by conventionalism, while the concept of convention increasingly tends to take the place occupied in positivism by that of experience. Whereas the final justification for knowledge according to positivism is supposed to be experience, the infinite regress which this tends to generate is abandoned by conventionalism, and the point at which the infinite search for justification is abandoned is marked by the concept of convention. This is true whether what is taken to be conventional are basic statements or methodological rules or both; in each case the designation of these posited elements as 'conventions' serves to withdraw them from the model of the justification of knowledge by deduction according to the logical schemata of implication, for conventions, according to conventionalism, cannot be derived from anything, they can only be arrived at by agreed decisions. This, however, is not a rejection of that model of justification, but only a decision to break the process off wherever it appears most convenient. It is significant that this refusal to continue with the positivist programme of final justification by experience is put forward, especially by critical rationalism, as anti-dogmatic: if scepticism towards any suggested basis for knowledge is always possible, then a final justification by such a basis would appear to rest on a dogmatic affirmation. Hence the attempt is abandoned, and conventions are (provisionally) introduced, which are themselves only justifiable by their fruitfulness, in terms of their consequences (Schnädelbach, 1971, pp. 150ff.; cf. Albert, 1968a, pp. 35ff.).

In addition, the shift from positivism to conventionalism is also marked by a move away from whatever remained of the epistemological claims of empiricism towards an emphasis on the logical analysis of systems of scientific propositions. This move can be seen in the development of logical positivism itself, for example in the writing of Rudolf Carnap: the rational reconstruction of scientific language-systems increasingly replaces the attempt to ground scientific knowledge in a basis of the given. The logic of science becomes less concerned with the origin of scientific concepts in a prior, given experience, and more concerned with the subsequent testing of judgements according to given rules of inference in a systematic context of justification. This shift is crystallized in the thesis, which becomes central to critical rationalism, of the separation of contexts of discovery from contexts of justification, of the origin (*Genesis*) of knowledge from its validity (*Geltung*), a thesis which can also be expressed as an abandonment of the psychologism of empiricist epistemology. However, just as the concept of convention is interpreted as the structural replacement for experience in empiricist epistemology, so Schnädelbach argues that the psychologistic model of empiricism is not entirely displaced, but rather re-interpreted conventionalistically (1971, pp. 152ff.).

It therefore becomes important to inquire into, not just the structural location of the concept of convention in the immanent development of the positivist problematic, but also the cognitive meaning of 'convention'. Conventionalism, after all, with its discovery that empiricist attempts to expunge all posited or produced elements of scientific knowledge were bound to fail, that empiricism could not justify itself empiricistically, appears to be a return to a kind of rationalism, an admittance of a priori knowledge. However, as its very name suggests, conventionalism does not, as Kant did, attach that a priori knowledge to a transcendental 'consciousness as such', but conceives it as a set of historically variable conventions (Schnädelbach, 1971, p. 135). It ought therefore to be possible, since transcendental questions are excluded, to ask empirical questions about the nature of these conventions: by whom are they made, and when, and under what conditions? However, argues Schnädelbach, conventionalists are reluctant to raise questions about the nature of conventions in this empirical form. Instead, they are represented as if they were almost some kind of epistemological 'Robinsonade', or, in a collective version, a kind of epistemological social contract. Popper himself, writing about words conceived of as 'conventional signs', suggests that logically considered, the meaning of a word

> is indeed established by an initial decision – something like a primary definition or convention, a kind of original social contract (1969, p. 18);

but the actual achievement of a consensus concerning such a convention appears to be a topic which conventionalism wants to treat as irrelevant (Schnädelbach, 1971, pp. 156f.).

We have seen that Popper, in *The Logic of Scientific Discovery*, had argued that the choice of conventions for methodological rules to characterize empirical natural science itself depended on the purposes that were set for science. Conventions were not a matter of pure decision, but were to be motivated by their adequacy in helping science fulfil its purpose. However, Popper did represent the choice of those purposes as itself a matter for pure decision. This rather curious proposal can be interpreted as an attempt to avoid an infinite regress, for if the purposes for science are themselves to be conceived of as conventions, it would be possible to ask what purposes those conventions were supposed to serve, and so on. Nonetheless, as Albrecht Wellmer has argued in a detailed study of Popper's work, rather than an infinite regress his conventionalism tends to result in a circular argument. In a number of places Popper makes it clear that technical applicability is an essential component of the aims of empirical natural science, and Wellmer (1967, p. 94) shows that the relationship of the conventional methodological rules to the aim of technical applicability cannot be adequately conceived of as one of means to ends, but rather as a redescription of the aims in methodological terms. Schnädelbach is led from this to the conclusion that both the pragmatic-technological aims and the methodological rules which are supposed to facilitate their achievement are in fact no more than redescriptions of the prevailing norms and methodologies of the empirical natural sciences as they actually exist. The freely chosen conventions which conventionalism posits turn out to be those which are as a matter of fact in force (1971, pp. 165ff.).

It would not be surprising if this were the case. What is more surprising is the impression given by conventionalism that the elbow-room available for considerable historical, or even personal, variation in the conventions on which empirical natural science could be built is rather wide. For this cannot be so. The demand for practical applicability puts severe constraints upon the range of conventions upon which an empirical natural science can be based. The elbow-room, and the free choice from an apparently wide range of alternatives, are equally illusory.

Nonetheless, the image of latitude is only projected in order that its narrowing down to a single set of conventions, for example those that Popper erects on the fundamental convention of the falsifiability criterion of demarcation, can be given philosophical justification rather than merely empirical currency. Hence to point to the congruence, if such it be, between the methodological rules proposed as conventions and the methodological rules which form the basis of science as it is actually practised, is taken by Schnädelbach as a demonstration that conventionalism in practice, and perhaps necessarily, reverts to what he calls a "positivism of the prevailing norms of science" (1971, p. 166). In other words, he claims that conventions which structurally take the place of a basis of experience in the problematic of positivism, take on the status of a new 'positive': the currently prevailing procedures of science do, despite conventionalist disclaimers such as Popper's rejection of a 'naturalistic methodology', come to be the indubitable basis on which the justification of scientific knowledge is to be, but now circularly, erected.

For Schnädelbach, this final twist in the history of positivism creates something markedly new. As we shall see in more detail below, a central feature of the way in which critical theory deals with philosophical issues, in this true to its Marxist origins, lies in its refusal ever to separate entirely philosophical problems or doctrines from the specific historical context in which they are situated. In the case of the history of positivism, Schnädelbach argues that the origins of the empiricist problematic, and in particular its sceptical motif, must be seen in relation to the historical conflict between the emerging empirical science of the seventeenth century and the traditional thought, both religious and metaphysical, to which science was a challenge. Empiricist philosophy was thus both theory of knowledge and critique of knowledge; its scepticism was not just a philosophical plaything but an element of enlightenment. The history of positivism, it is argued, continued to combine these elements: the philosophical justification of scientific knowledge, and the scepticism directed towards any proposed basis, was at the same time a part of the offensive legitimation of scientific knowledge against a defensive but still powerful tradition. Even logical positivism, he claims, was still a knowledge-critical philosophy in this sense. But the conventionalism most strongly represented by critical rationalism has abandoned those knowledge-critical pretensions precisely because the traditional forms of knowledge, religion and metaphysics, are no longer strong enough to constitute an obstacle. Science then no longer needs philosophical justification, since it no longer needs social legitimation. It is its own legitimation, and the philosophy which claims to represent it can relapse into a positivism in which the affirmative element, so far as the norms of science are concerned, sweeps away the sceptical one.

We are now in a position to summarize the identification of positivism advanced by critical theory in the context of the philosophy of the natural sciences. It is clearly more complex than the account of positivism from which critical rationalism is at pains to distance itself. It is advanced as a dynamic conception of a philosophical problematic, in which a small number of initial epistemological and ontological presuppositions generate a range of problems which form the motor for the further development of this epistemological tradition. The positivist search for an indubitable experiential basis for scientific knowledge has an inherent tendency towards conventionalism, and the inadequacy of the concept of convention generates an inherent tendency in conventionalism to return to a new form of positivism, in which the indubitable basis is the existing structure of science itself. Thus both that which has identified itself as positivist and that which has represented itself as antipositivist or post-positivist are equally identified with 'positivism' in this wider sense, so long as neither break through the restrictions imposed by the initial empiricist presuppositions.

3.2. Critical Theory's Critique of Positivism: The Kantian Background

It must be stressed at the outset that critical theory does not deny that positivism is capable of providing an adequate analysis of the methodology of the natural sciences. On the contrary, those who have written in detail on positivism and critical rationalism such as Wellmer and Schnädelbach, are quite explicit that the failings of positivism, as they see them, do not lie at the level of descriptive methodology (Wellmer, 1967, p. 100). Critical theory's critique of positivism is motivated, not by a deep interest in the logic of the natural sciences, but by concern for the consequences of positivism's universalization of that logic for the human or social sciences. It might therefore be thought that an examination of critical theory's critique of positivism should be postponed, and should find its place in the study of forms of antipositivism in the philosophy of the social sciences – that it would appear as a critique of the typical 'central tenet' of the unity of science, usually taken as one defining characteristic of positivism.

However, if the problem is to be raised in the form in which Wellmer sets it, or in some form which poses the question whether or not it is the case that the methodology of the natural sciences, as described by, say, critical rationalism, is equally applicable to the study of human society, then it is plausible to argue that the issue is best treated at a more general level, at which it can also be asked what it is that makes the methodology of the natural sciences as they have developed appropriate to the study of nature, whether or not they are equally appropriate to the study of human society. In fact, to raise the problem in this way is essential to an understanding of critical theory's critique of positivism, for it is a vital element of that critique that positivism has regressed from the level of epistemology to that of the logic of science, and cannot provide an answer to this question, not least because it does not ask it. In this respect, critical rationalism is regarded as an advance over earlier positivism, but not a decisive one; as Habermas puts it:

> neither the conventionalist nor the naturalistic interpretation does justice to the choice of methodological rules (Adorno *et al.*, 1976, p. 212).

The model of epistemology, in comparison with which modern philosophy or logic of science is judged to be a regression, is taken from Kant's 'critical philosophy'. In order to understand how critical theory uses Kant's philosophy as a weapon of criticism, although it also becomes an object of criticism, it will be useful to present, as a preliminary, a brief account of Kantian epistemology.

Kant's philosophy was based on a view of science different from that of empiricism, a view which was articulated at two levels. Firstly, at the level of empirical science itself, Kant claimed that a pure receptivity model of scientific knowledge was simply an inadequate account of how science, over the century and a half before Kant himself wrote, had made such rapid strides. A passive model of knowledge could not do justice to scientific progress, an active model was necessary instead:

> students of nature ... learned that reason has insight only into that which it produces after a plan of its own, and that it must not allow itself to be kept, as it were, in nature's leading-strings, but must itself show the way with principles of judgment based upon fixed laws, constraining nature to give answer to questions of reason's own determining. Accidental observations, made in obedience to no previously thought-out plan, can never be made to yield a necessary law, which alone reason is concerned to discover (Kant, 1929, p. 20).

Only when the investigation of nature is based on such a plan will the progress of science be placed on a secure foundation.

There are therefore necessarily 'produced' or a priori elements involved in all empirical scientific knowledge. If science is to be understood philosophically and given epistemological foundations, philosophy will have to be able to account for these a priori elements. However, traditional metaphysics, which claimed this task for itself, was rejected by Kant on the grounds that it produced speculative systems of ideas which could only be propounded or attacked dogmatically. In the spirit of 'the age of criticism', metaphysics itself was to be put on a scientific basis. Here the second level of articulation of Kant's new view of science came into play. For if empirical natural science could only be understood on the assumption that reason brought with it a priori categories to the observation of nature, should not metaphysics, if it is to be scientific, follow the same procedure. Kant's Copernican Revolution, that is, involved the shift from a conception of metaphysics as the study of those objects which lay beyond experience, to one which took the object of metaphysics to be the powers of human reason itself. Thus an understanding of the 'produced', the a priori elements which enter into empirical natural science, can only come about through a process of reflection, in which reason reflects its own powers or faculties. What in positivism can only be understood as conventional, is in Kant a priori knowledge. This reflective model of critical philosophy is taken over from Kant by critical theory, even if the latter ends by drawing very different conclusions from it.

Before we come to critical theory's divergence from its Kantian heritage, however, it is worth sketching out further the main elements of Kant's own philosophy, just to the extent that it can serve as a base-line. Firstly, as we have seen, the passive model of cognition found in empiricism (which Kant calls 'scepticism') is replaced by an active model. This is achieved by distinguishing two faculties of the knowing subject: the sensibility and the understanding. The sensibility is taken over from empiricism as the faculty of receptivity to sense-data. The understanding is conceived of as the faculty which imposes order on the mass of sensations received by the sensibility. The sensibility is presented with a manifold of representations, which must all conform to the a priori forms of appearance of space and time, but which are otherwise unstructured. As such they cannot be known by the knowing subject: they must have structure imposed upon them. This task is performed by the understanding, which connects together representations into judgments about objects; the activity of the understanding is known as synthetic activity, or synthesis. The a priori rules, according to which this synthetic activity is performed, are the twelve categories of the understanding, which are derived from the forms of judgments in formal logic.

The forms of sensibility and the categories of the understanding could be called 'subjective', in that they are forms of human experience. But Kant is concerned to show that they are at the same time 'objective', in that they are not psychological elements subject to personal variation, but necessarily the same for all human beings. The reconciliation of an 'activist' theory of knowledge with the objectivity of knowledge is indeed one of Kant's central concerns. Objectivity is conceived of as necessary agreement between human beings, as a necessary minimum of shared presuppositions and of community that make understanding and dialogue possible (Goldmann, 1971, p. 152; Beck, 1963, p. 22).

Through its synthetic activity, the knowing subject constitutes out of the manifold of its representations a domain of objects to which the categories are applicable, a domain of phenomenal objects of experience. Of such objects man can have knowledge, but the objects of empirical intuition are themselves 'undetermined', partial appearances of things as they are (Kant, 1929, p. 65). The completely determined things as they are cannot be known to the understanding, for the sensibility has no received information to present to the understanding for synthesis, and both are necessary to knowledge: "thoughts without content are empty, intuitions without concepts are blind" (1929, p. 93). But Kant states that the human mind has a third faculty, reason, which can entertain ideas which go beyond experiential knowledge. Such an idea is the completely determined thing in itself:

> to know a thing completely, we must know every possible predicate, and must determine it thereby, either affirmatively or negatively. The complete determination is thus a concept, which, in its totality, can never be exhibited *in concreto*. It is based upon an idea, which has its seat solely in the faculty of reason (1929, p. 489).

Traditional metaphysics had assumed that, since empirical knowledge of nature was necessarily partial, speculative reason must go beyond or outside knowledge of nature to find the answers to questions about how things really are, as opposed to how they appear to human consciousness. But Kant argued that there can be no knowledge of things as they really are, since in going beyond the phenomena reason would sever the link that tied knowledge to the world of experience. It would continue to use concepts, namely the categories, but in an empty, contentless way. Also, in attempting to apply the categories to a domain of objects not related in any way to the manifold of representations, reason entangled itself in pairs of contradictory, but equally provable, statements, which Kant termed antinomies. The demonstration of the antinomies was designed to prove that reason strayed too far afield if it tried to push knowledge beyond the phenomena.

However, although objects beyond the realm of the sensible could not, on this argument, be known, they could be thought. The concepts of objects beyond the sensibility correspond to what Kant termed the Ideas of reason. Thinking of such concepts was, nevertheless, not pure fantasy and a waste of time, but rather absolutely essential to the process of extending knowledge. For if knowledge is to be extended, it is necessary for the knowing subject to have an idea of what is not yet known, in order to know in which direction the incompleteness of its present knowledge is to be, even if only partially, remedied. As Kant put it, reason is

> the faculty which prescribes to the understanding the rule of its complete employment (1929, p. 489).

In other words, the ideas of reason, while they postulated objects which according to Kant could never be known, were necessary because they acted as regulative principles, or methodological rules, to use the terminology of the earlier argument, which directed the pursuit of progress in scientific knowledge of nature. The idea of things in themselves, things as they really are, the unconditioned, completely determined totality, acts as an unattainable goal to which the process of expanding knowledge must be oriented.

To use the transcendental ideas of reason to refer to objects, which might be considered to exist and be capable of being known in the same way as the phenomenal world can be known to the understanding, would be therefore to go beyond the limits necessarily set to human knowledge. The negative critique of the speculative metaphysics that is generated by such transgression also has a positive lesson:

> it likewise teaches us . . . that human reason has a natural tendency to transgress these limits, and that transcendental ideas are just as natural to it as the categories are to understanding (1929, p. 532).

The immanent, regulative use of the transcendental ideas of reason is thus the controlled and disciplined expression of the natural tendency to seek the transcendental itself. In epistemology this tendency expresses itself as a search for the complete determination of the given, knowledge of the totality, of the thing in itself. But this is only the epistemological side of what for Kant is a philosophical view of man, a philosophical anthropology, according to which man's destiny is to strive towards the unconditioned. If we ask where the idea of this philosophical anthropology itself comes from, the answer is to be found in the basic argument of the critical philosophy as it was outlined above: it is an idea that is known a priori, since, according to Kant's Copernican revolution, it comes from the human mind itself rather than being supplied to it from outside. It is itself a transcendental idea, and acts as a practical postulate for both epistemology and morality.

Thus, whereas on the side of the object of knowledge this striving towards the absolute appears as a search for the thing in itself, the objective totality, Kant's philosophical anthropology does not have a static and eternal conception of the knowing subject, mankind, but an idea of an absolute towards which mankind can also strive. Just as the forms of the intuition and the categories of the understanding constitute necessary minimum conditions for a dialogue between human beings that can lead to objective, that is universal and necessary, knowledge, so reason can have an idea of the maximum conditions for such a dialogue, the idea of a perfect human community. The idea of a better human community, a society of citizens of the world which will enjoy perpetual peace and perfect justice, is clearly expressed in Kant's writings, especially those on politics and history (Kant, 1963). It is a transcendental idea, an idea of a subjective totality, and because Kant does not believe it is capable of historical realization, but that it nonetheless remains necessary as a practical postulate to guide action (just as the transcendental idea of the thing in itself is necessary as a guide to the progress of scientific knowledge), it is an idea that is not expressed as a philosophy of history but rather transferred to the philosophy of religion. Since people cannot be expected to orient their action towards an absolute goal that is in principle unrealizable, and since Kant believes the perfect human community is on the one hand a necessary idea but on the other hand historically unrealizable, the hope for the future is removed to a suprasensible realm of the kingdom of God and the immortality of the soul. As transcendental ideas, these refer not to subjects that can be known, but to something that can be hoped for:

> I have therefore found it necessary to deny *knowledge*, in order to make room for *faith* (1929, p. 29).

These, then, are the leading themes of Kant's own implementation of his newly discovered method of critical reflection. In this method, in which the knowing subject reflects back to itself its own powers of reason, the aim is to produce a systematic account of all of mankind's diverse modes of knowledge in relation to each other, specifying the limits and functions of each mode in relation to the whole. This is what Kant referred to as "architechtonic" (1929, pp. 653ff.). At the basis of this architechtonic is a conception of man, a philosophical anthropology, whose 'whole vocation' constitutes the essential end to which all the various modes of knowledge are to be related.

The critique of pure reason is a part of this entire system, the principle task of which, as Goldmann expresses it,

> consists in the struggle against two dangerous illusions which could lead man to betray his destiny and to abandon the search for the absolute, namely:
>
> (a) *the transcendental employment of the categories*, the idea that the human faculty of knowledge, such as it is and without qualitative change, can attain the absolute, the illusion of all dogmatic metaphysics, and
>
> (b) *sceptical empiricism*, the contrary affirmation, according to which the unconditioned, the totality in general, is unreal and inaccessible to any kind of knowledge. If this were the case, the aspiration towards a higher state would become meaningless: speculative ideas would lose their regulative meaning and practical postulates their practical meaning (1971, pp. 163f.).

This is the core of Kant's critique of what later became known as 'positivism', and which was taken over and extended by critical theory. In general terms, relative to Kant's critical philosophy, positivism is condemned as 'pre-critical' (Schnädelbach, 1971, p. 129).

While Kant was just as impressed with Newtonian mechanics as later the logical positivists were with Einsteinian relativity theory, and like them took the object of his admiration as a model for natural science in general, he did not identify that natural science with knowledge as such, but located it in its own proper place within the overall system, or architechtonic, of human knowledge. Thus, whereas positivism set itself the task of justifying scientific knowledge purely in terms of its adequacy to its object and hence, because of its receptivity model of cognition, in terms of its indubitable basis in experience, Kant could seek a justification for scientific knowledge of nature in terms of its contribution to the ends of human reason.

In view of positivism's inability to execute its own programme and its subsequent dissolution into conventionalism, the Kantian critique might appear an adequate basis for critical theory's rejection of classical positivism. The case of conventionalism itself, especially in the form of critical rationalism, is more complicated, all the more so since from its inception in Popper's

Logic of Scientific Discovery, critical rationalism itself claimed to be continuing in the Kantian tradition, designating the problem of demarcation "Kant's problem" (1968, p. 34). Popper did indeed originally accord to certain methodological rules, such as the "principle of causality", a regulative function not unlike that which Kant proposed for transcendental ideas (1968, p. 61). The essential difference, however, is that whereas Kant claimed to derive such ideas by reflection, Popper wanted to postulate them as conventions introduced by decision.

Later, despite an occasional tendency to interpret Kant as a sceptical empiricist (1969, pp. 179f.), Popper came to see Kant's stress on a priori elements in knowledge as a theory of unchangeable inborn expectations by which we impose laws on nature. Thus, although Kant recognized that the process of gaining knowledge involved an 'active' rather than a 'passive' subject, this was a 'conservative' rather than a 'revolutionary' activism, and according to critical rationalism Kant had no real conception of scientific progress (Lakatos and Musgrave, 1970, p. 184).

Popper's response to this was to cast his theory of scientific progress in evolutionary terms (cf. Campbell, 1974). The active component of scientific expectations was indeed a set of inborn expectations, but of a very general kind, namely "the expectation of finding a regularity", which was "connected with an inborn propensity to look out for regularities, or with a *need* to *find* regularities" (1969, p. 47). The a priori is thus a 'psychological' one. This theory attributes to organisms a learning mechanism, by which they can improve their reactions to their environment and thus their survival chances. Humans follow a similar learning process, except that they have externalized their expectations into a "third world", as Popper later called it, so that evolutionary selection works on theories instead of on organisms themselves; "scientists try to eliminate their false theories, they try to let them die in their stead" (Popper, 1972, p. 122).

Briefly put, this is Popper's theory of the a priori elements in knowledge, his response to the Kantian tradition. Central to it is the analogy between scientific progress and biological evolution. From the standpoint of critical theory, Albrecht Wellmer has subjected this biological analogy and the use Popper makes of it to detailed examination. There are two strands to his criticism. Firstly he argues that Popper obliterates an essential distinction between a disposition to react on the part of an organism, and a linguistically formulated expectation on the part of a human being. The first can be formulated in terms of stimulus and response, and to attribute a disposition to react to an organism is itself to advance an hypothesis which is open to experimental refutation (Wellmer, 1967, p. 55). The latter is more like what Peirce termed a 'practical belief', a belief about the likely outcome of certain courses of action, which provides people with information on the basis of which they can make more or less rational plans concerning the best means of attaining their ends. The 'expectation' that a certain powder is fatal when consumed, for example, is not a disposition to react but rather a belief on the basis of which I can react rationally to specific situations, depending on whether I want to stay alive, to protect another person, to kill an enemy, or commit suicide (Wellmer, 1967, p. 55). If this distinction is made, it is no longer acceptable to assimilate scientific theory to the reaction-dispositions of organisms, which implies that Popper's attempt to provide an a priori for science through biological analogy must collapse. It is thus misleading to suggest, as Popper does, that in the history of science "there is only one step from the amoeba to Einstein" (1972, p. 347).

Secondly, however, Wellmer interprets Popper's biological considerations, and the evolutionary epistemology to which they lead, as evidence of "a transcendental reflection which has been broken off" (1967, p. 65). In other words, the attempt to generate an a priori for scientific knowledge by means of the biological analogy is, in partial form, a theory of the nature of the knowing subject and an account of its cognitive powers. It is a partial theory, in comparison with that of Kant, because it fails to conceive itself as a process by which the knowing subject reflects its own powers of reason, but rather represents itself as itself an objective scientific theory, the theory of evolution. And its result is, according to Wellmer, the reduction of epistemology to methodology, or to the logic of science, which, as we have seen, is the central theme of critical theory's critique of positivism. Wellmer (1967, p. 65) quotes Popper as follows:

> Assume that we have deliberately made it our task to live in this unknown world of ours; to adjust ourselves to it as well as we can; to take advantage of the opportunities we can find in it; and to explain it, *if* possible (we need not assume that it is), and as far as possible, with the help of laws and explanatory theories. *If we have made this our task, then there is no more rational procedure than the method of trial and error – of conjecture and refutation*: of boldly proposing theories; of trying our best to show

that these are erroneous; and of accepting them tentatively if our critical efforts are unsuccessful (Popper, 1969, p. 51).

Here in this passage we find *both* the predominant place accorded to methodology in Popper's theory of knowledge *and* the partial theory of the knowing subject: it is because 'we' are as we are that the method of trial and error is appropriate.

It is curious, in the light of the polemical nature of the *Positivismusstreit*, that, by comparison with the 'positivists' whom Popper himself criticizes, Popper and critical rationalism are in fact accorded considerable philosophical respect by critical theorists. The reason for this is, as Wellmer puts it, that Popper's critique of positivism is, in its approach, a transcendental critique, even though Popper misinterprets Kant, and even though he does not always recognize the transcendental nature of his critique of positivism. In fact, in one respect Popper goes beyond Kant and hits a weak point in the latter's transcendental philosophy:

This weakness, however, is not to be found in the idea, which Popper advances, that Kant failed to distinguish between the fact of a logical-psychological a priori and the validity of that a priori; rather was it the case – and on this point, it seems to me, both Popper and Hume are in the right against him – that Kant, in his account of the synthetic activity of the transcendental subject, suppressed any reference to the real activity and needs of empirical subjects (Wellmer, 1967, p. 66).

For while critical theory's recourse to Kant does make clear the central thrust of its critique of positivism, to leave matters there would be most misleading. Before concluding this account of critical theory's critique of positivism in the philosophy of the natural sciences, therefore, it is essential to set out the framework within which a more positive contribution to the epistemology of the natural sciences may be made by critical theory, which, as the previous citation from Wellmer suggests, develops in part from a critique of the *purely* transcendental nature of Kant's own philosophy.

3.3. Critical Theory and the Natural Sciences: Beyond Kant and Positivism

For Kant, as we have seen, knowledge cannot be conceived as direct acquaintance with an independently structured reality. On the contrary, knowledge is only possible if the knowing subject constitutes out of the manifold of its representations a world of ordered experience. Once knowledge is conceived as a mediation between the object of knowledge and the knowing subject, it becomes necessary to provide a theory which will account for this activity of constitution, or synthetic activity, in terms of the nature of the knowing subject. Kant was seeking a theory which would make the ways in which this world of ordered experience was constituted a priori necessary. Hence he could not locate synthetic activity in the consciousness of empirical individuals, since then nothing would guarantee that different individuals would constitute the world in the same way, or even that one individual would constitute it in the same way from one moment to another. Instead, he had to look for a transcendental ground which would guarantee synthetic unity, and he found it in the notion of a "pure original unchanging consciousness" which he named "transcendental apperception" (Kant, 1929, p. 136). It is to this transcendental consciousness, to this infinite subject, that the forms are innate, the forms which structure the unstructured contents of consciousness and thus make knowledge possible. These forms are then elaborated as the categories of the understanding and the transcendental ideas of reason.

Kant's epistemology gives rise to characteristic problems about the nature of this transcendental consciousness. Thus Alan Montefiore puts in rather more straightforward terms the difficulty revealed by Wellmer:

> there is clearly something messy and unstable about a theory in which the very possibility of objective experience, and that of a self-conscious subject, are understood as dependent upon each other, if it cannot, at the same time, provide a coherent account of the nature of that subject in terms of its relation to 'real life' individuals or groups (Montefiore, 1976, p. 673).

Or the problem can be put in a different way: while on the one hand Kant grounds synthetic unity in a 'pure original unchanging consciousness' which is conceived in purely individualistic terms of the 'I think' which accompanies all my representations (1929, p. 152), he also, as we have seen, admits as a transcendental idea a concept of a subjective totality, a perfect human community characterized by perfect understanding between its members. The synthetic unity supposedly based in original consciousness can guarantee only formal agreement, which is debarred from knowledge of the thing in itself; but there appears to be some connection, which remains unclear in

Kant, between the transcendental idea of the thing in itself, the objective totality, and the transcendental idea of a perfect community, the subjective totality (Goldmann, 1971, pp. 162f.). This suggests that, rather than the original consciousness being unchanging, there is the possibility of historical development towards totality, and again the problem arises of the relationship between this historical development as a purely abstract philosophical idea and 'real life' historical development.

Because of these problems, critical theory's reintroduction of the problem of constitution, or of synthetic activity, into a philosophy of science which is diagnosed as having forgotten that problem cannot just be a return to Kant. The difficulties, which are themselves highlighted by the partial transcendental reflection of critical rationalism, have to be acknowledged and treated. The lines of argument with which critical theory comes to terms with these difficulties are, as is well known, derived from Hegel and Marx, and more recently also from the pragmaticist philosophy of Peirce. It is necessary here to provide only a brief outline of these arguments.[6]

Firstly, following Hegel, the ahistorical nature of Kant's transcendental consciousness is abandoned.

> Kant proceeds from the identity of the 'I' as an original unity of transcendental consciousness. In contrast to this, Hegel's fundamental experience of the 'I' as an identity of the universal and the singular has led him to the insight that the identity of self-consciousness is not an original one, but can only be conceived as one that has developed (*geworden*) . . . The radicalization of the critique of knowledge carried out as a science of appearing (*erscheinende* = phenomenal) consciousness consists precisely in relinquishing the viewpoint of a 'ready-made' or 'completed' subject of knowledge (Habermas, 1974, pp. 156f.).

The historical development of the subject of knowledge is thus conceived as a *self-formative* process. This theme is developed in the form in which it is found in Marx, who in the philosophy of history replaced Hegel's infinite Spirit with 'the historical life-process of finite men' (Schmidt, 1971, p. 108). The possibility of synthetic unity is derived from the common membership of these finite men in the human species: thus Habermas proposes the thesis that

> the achievements of the transcendental subject have their basis in the natural history of the human species (1972, p. 312; cf. Horkheimer, 1972, pp. 202ff.).

Secondly, critical theory seeks to overcome the Kantian separation between pure and practical reason, between theory and practice. In this it follows Marx's critique of Hegel in the final part of the *Economic and Philosophical Manuscripts*, where Hegel is first praised for having conceived of the self-creation of man as a process which man brings about through his own labour, and then criticized for being able only to recognize abstract mental labour as truly human, and not real productive labour (Marx, 1975, p. 386). This theme was taken up again by Marx in his 'theses on Feuerbach', in which Feuerbach's materialism is condemned as just as contemplative as all earlier forms of materialism. Feuerbach, it is said,

> wants sensuous objects, really distinct from thought objects, but he does not conceive human activity itself as *objective* activity. Hence, in *Das Wesen des Christentums*, he regards the theoretical attitude as the only genuinely human attitude, while practice is conceived and fixed only in its dirty-judaical manifestation (Marx, 1975, pp. 421f.; cf. Bloch, 1971, pp. 64ff.).

This criticism is further developed in *The German Ideology*, and set in the context of the theory of historical materialism, which is stated in a coherent form for the first time in that work. The basic premise of historical materialism is the unique position of mankind in its natural environment:

> Men can be distinguished from animals by consciousness, by religion or anything else you like. They themselves begin to distinguish themselves from animals as soon as they begin to *produce* their means of subsistence, a step which is conditioned by their physical organization. By producing their means of subsistence men are indirectly producing their actual material life (Marx and Engels, 1968, p. 31).

The human species, like all other species, exists in a material, or natural, environment. Unlike all other species, Marx and Engels claimed, the human species makes a break away from mere adaptation to a material environment that is taken as given and unalterable, and instead develops the capability of working on that environment and transforming it. Hence mankind's relationship to nature is not an abstract, theoretical, and contemplative one, as is assumed by Kantian and idealist epistemology and by Feuerbach, but a practical and transformative one. This is itself not purely an anthropological or historical thesis, but also an epistemological presupposition. For it implies that mankind's knowledge of the natural world is guided and directed by its practical relationship to that world. This practical relationship generates a transcendental framework in which the world of nature may be constituted and known. At first this knowledge remains at a relatively primitive level, corresponding to an early stage in the development of the productive forces. As mankind gradually proceeded to improve its understanding of natural processes, the systematic natural sciences developed as a continuation of such pre-scientific knowledge, but natural science remains bound to the practical relationship of man to nature, however complex and 'theoretical' it might become. Thus Marx and Engels, continuing their critique of Feuerbach, note that the latter

> speaks in particular of the perception of natural science; he mentions secrets which are disclosed only to the eye of the physicist and chemist; but where would natural science be without industry and commerce? Even this 'pure' natural science is provided with an aim, as with its material, only through trade and industry, through the sensuous activity of men (1968, p. 58).

And already in the *Manuscripts* Marx had stated that

> industry is the *real* historical relationship of nature, and hence of natural science, to man (1975, p. 355).

In his programmatic essay, 'Traditional and critical theory', which set out the basic framework within which members of the Institute for Social Research at Frankfurt had been and were still operating, Max Horkheimer took over this practical epistemology of the natural sciences from Marx and contrasted it with the traditional, or bourgeois, conception of scientific activity which, even where it recognized the practical applicability of knowledge as an important element in the natural sciences, tended to see this exclusively as a separate kind of activity, the subsequent application in the form of technology of what was essentially pure knowledge, quite independent of the productive organization of society (1972, pp. 196ff.). Critical theory, unlike traditional theory, is able to grasp the essential unity of theory and practice. In addition, however, Horkheimer also stresses another aspect of the Hegelian historicization of the knowing subject in combination with Marx's materialism: if, on the one hand, synthetic activity is not to be conceived as performed by an ahistorical, unchanging consciousness but by an historical subject engaged upon a self-formative process and if, on the other hand, this historical subject does not have a merely theoretical but a practical relationship to nature which is manifested in productive labour, then it appears logical to conclude that the nature of that synthetic activity, and hence the form of natural science, will change as the subject's, namely mankind's, productive relationship to nature itself is transformed. It seems possible to argue that change in the structure of natural science, both at the level of the particular scientific problems that will be researched, but also at the level of the fundamental structure of scientific explanation, is dependent on change in the social organization of production (Horkheimer, 1972, pp. 194ff.).

The way was then open for two different strands of thought within critical theory concerning the epistemology of the natural sciences. One strand, which may be derived from Horkheimer, and represented both by Herbert Marcuse and by an early associate of the Frankfurt School, Alfred Sohn-Rethel, stresses the historical transformation of basic structural features of the natural sciences, a transformation which is not conceived as having reached its conclusion. The other strand, most clearly represented by Jürgen Habermas, stresses features of the practical, or as he terms it 'technical', relationship of man to nature which are conceived of as universally necessary and hence not subject to historical variability. It will be useful to outline in a little more detail these two strands of thought, in the light of both the earlier problem raised by conventionalism, of the apparent historical variability of methodological rules underlying scientific activity, and of the discussion still to come, namely the scientific realist critique of positivism, which will also raise the problem of variation of the methodological rules of natural science.

3.3.1. The Critique of Technical Reason

The first strand of thought is perhaps most typical of the early writings of the Frankfurt School, and of the shift to the 'critique of instrumental rationality' signalled, as we have seen, by Horkheimer and Adorno's *Dialectic of Enlightenment*. This work represents an abandonment of Marx's optimism, which he shared with the enlightenment as a whole, that the progressive development of the productive forces would make possible the rationalization of the relations of production, and the bringing under human control of all aspects of social life. Instead, Horkheimer and Adorno saw in that rationalization a complete reification of individuals and of social relationships; the process of the domination of nature carried with it not the potentiality for freedom but rather the subjection of human beings under forms of social organization made necessary by precisely that same process.

It is possible to derive from this critique of instrumental reason a two-pronged renewal of critical social theory (see Kreckel, 1975, pp. 89ff.). The first of these is the 'critique of technocracy', the thesis that

> the growth of economic productivity . . . allows the technical apparatus and the social groups which administer it a disproportionate superiority over the rest of the population (Horkheimer and Adorno, 1973, p. xiv).

From this can be developed the argument, with which we will be concerned below, that in advanced industrial societies technological rationality increasingly takes on the function of an ideology to legitimate domination, which penetrates all areas of society. The second is the 'critique of positivism', this time in the more restricted sense of an invasion of the methods of natural science and hence of technological rationality into the social sciences. But there is also a third thesis which can be drawn from the critique of instrumental rationality which is more directly of concern at this point: that the natural sciences themselves contribute directly to the domination of men by men, and that a social transformation in the direction of greater freedom and equality would require a transformation of the natural sciences themselves. This thesis is advanced by Marcuse:

> science, *by virtue of its own method* and concepts, has projected and promoted a universe in which the domination of nature has remained linked to the domination of man – a link which tends to be fatal to this universe as a whole. Nature, scientifically comprehended and mastered, reappears in the technical apparatus of production and destruction which sustains and improves the life of the individuals while subordinating them to the masters of the apparatus. Thus the rational hierarchy merges with the social one. If this is the case, then the change in the direction of progress, which might sever this fatal link, would also affect the very structure of science – the scientific project. Its hypotheses, without losing their rational character, would develop in an essentially different experimental context (that of a pacified world); consequently, science would arrive at essentially different concepts of nature and establish essentially different facts (1968, pp. 135f.).

Not altogether surprisingly, Marcuse seems to have no very clear idea of what this 'alternative science' would look like; indeed, elsewhere in his work he appears to retract the suggestion, or weaken it to the apparently more plausible one of a changed social organization of and control over science and technology (1972b; cf. Leiss, 1972, 1975). However, the underlying idea deserves further consideration, the idea that structural elements of natural scientific knowledge change as people enter into new productive relationships with each other and with physical nature.

This idea is already present in embryo in Marx. As we have seen, Marx's epistemology is one in which the transcendental subject is interpreted materialistically as the human species in the process of self-formation. Because it exists in a natural environment and develops the capacity to work on and transform that environment, mankind's synthetic activity in so far as it concerns the constitution of the world of nature is also interpreted materialistically as the activity of productive labour. Productive labour itself, however, undergoes transformation throughout history. By emphasizing this historical transformation, the characteristic difference between Kant and Marx, as Habermas puts it, becomes clear:

> The synthesis of the material of intuition by the imagination receives its necessary unity through categories of the understanding. As pure concepts of the understanding these transcendental rules of synthesis make up the internal and unalterable inventory of consciousness as such. The synthesis of the material of labour by labour power receives its actual unity through categories of man's manipulations. As instruments in the broadest sense, these technical rules of synthesis take on sensuous existence and belong to the historically alterable inventory of societies (1972, p. 35).

But labour is not just a matter of technical manipulation; it is also necessarily incorporated into forms of social organization. In all class societies, all societies characterized by a rigid division of labour, these forms of social organization are, according to Marx, alienated forms, and the labour performed within them alienated labour (Marx, 1975, pp. 322ff.). The epistemological consequences of this are, firstly, that it might be expected that the forms under which nature is known would change as both technical and organizational aspects of productive labour change, and secondly, that the forms under which nature is known may be expected to partake in some way of the alienation of that productive activity, and in the future to share in the abolition of alienation.

Marx's concept of nature has been subjected to detailed examination by Alfred Schmidt in his work of that name (Schmidt, 1971), and it is unnecessary as well as impossible to recall his analysis here, save for one or two points. Schmidt argues that, for Marx, the changing relationship of man to nature, of Subject to Object, manifested itself in a shift from a view of nature which emphasized objective features to one which emphasized subjective features.

One can say in general of the historico-economic process of the transformation of Object into Subject and of Subject into Object, that under pre-industrial conditions the objective, natural moment is dominant, whilst in industrial society the moment of subjective intervention asserts itself in increasing measure over the material provided by nature (Schmidt, 1971, p. 121).

Thus

in an agrarian economy, men take up a passively receptive attitude to nature, which appears directly as wealth in the means of subsistence: "Land is still regarded here as something which exists naturally and independently of man, and not yet as capital, i.e. as a factor of labour. On the contrary, labour appears to be a factor of nature" (1971, p. 121; cf. Marx, 1975, p. 344).

In an industrial society nature is increasingly seen from the subjective standpoint of its transformability by labour:

Thus capital creates the bourgeois society, and the universal appropriation of nature as well as of the social bond itself by the members of society. Hence the great civilizing influence of capital; its production of a stage of society in comparison to which all earlier ones appear as mere *local developments* of humanity and as *nature-idolatry*. For the first time, nature becomes purely an object for humankind, purely a matter of utility (Marx, 1973, pp. 409f.).

The epistemological aspect of this, Schmidt suggests, is the transition from early materialism, which "viewed nature as a fixed datum, and knowledge as the mirror which reflected it" (1971, p. 122), to the materialism of Marx's own 'Theses on Feuerbach', stressing the role of practical activity in the cognitive process. Thus, in industrial society, nature

ceases to be recognized as a power for itself, and the theoretical discovery of its autonomous laws appears merely as a ruse so as to subjugate it under human needs, whether as an object of consumption or as a means of production (Marx, 1973, p. 410).

However, since Marx was of course never primarily concerned with the epistemological aspects of historical materialism, this remains no more than a hint towards a more adequate account.

Secondly, and even more sketchily, it is possible to find remarks in Marx which suggest that he did believe that the natural sciences were affected by the alienated form of productive labour which functioned as synthetic activity to constitute their object-realm. The clearest example is contained in the following passage:

> The *natural sciences* have been prolifically active and have gathered together an ever growing mass of material. But philosophy has remained just as alien to them as they have remained alien to philosophy. Their momentary union was only a *fantastic illusion*. The will was there, but not the means. Even historiography only incidentally takes account of natural science, which it sees as contributing to enlightenment, utility and a few great discoveries. But natural science has intervened in and transformed human life all the more *practically* through industry and has prepared the conditions for human emancipation, however much its immediate effect was to complete the process of dehumanization. *Industry* is the *real* historical relationship of nature, and hence of natural science, to man. If it is then conceived as the *exoteric* revelation of man's *essential powers*, the *human* essence of nature or the *natural* essence of man can also be understood. Hence natural science will lose its abstractly material, or rather idealist, orientation and become the basis of a *human* science, just as it has already become – though in an *estranged* form – the basis of actual human life. The idea of *one* basis for life and another for *science* is from the very outset a lie. Nature as it comes into being in human history – in the act of creation of human society – is the *true* nature of man; hence nature as it comes into being through industry, though in an *estranged* form, is true *anthropological* nature (1975, p. 355; cf. 1976, pp. 493f.).

There is always the danger, with these early manuscripts, of 'over-interpretation' of rather vague remarks, although in this case it may be significant that a similar idea is expressed in the later *Capital*. Marx does seem to draw a distinction between an alienated and a non-alienated form of the natural sciences, corresponding to alienated and non-alienated forms of industrial production. But it remains quite unclear whether this difference is supposed to be one of fundamental categories, in something like a Kantian sense, or one of perhaps less fundamental aspects of the social organization of science. Thus Istvan Meszaros, in a commentary on this passage, sees the alienation of natural science as the fragmentation of the particular sciences, and the unplanned, unconscious nature of their development as corresponding to the structure of alienated productive activity in general (1970, p. 102). This fragmentation could, however, be a purely sociological feature, with no epistemological significance.

It must be uncertain, therefore, whether and in what sense Marx should be seen as Marcuse's forerunner in setting out the idea of an 'essentially different' natural science. It is much less uncertain, however, that Marx did develop the idea that the synthetic activity which makes natural scientific knowledge possible is not the activity of an unchanging original consciousness, but the productive activity of human beings in their encounter with their natural environment. The historical variation in synthetic activity which this implies is developed by Marx only to a limited degree; the study of this historical variation is taken up in more detail from Marx by writers in this first strand of critical theory.

The most far-reaching attempt to derive categories of thought by means of which nature may be understood from the social organization of productive labour has been conducted by Alfred Sohn-Rethel, who for present purposes may be treated, as he considers himself 'in a sense', as part of the Frankfurt School (Slater, 1977, pp. 83ff.). Sohn-Rethel explicitly took up the problem of the categories as a problem of the reinterpretation of the Kantian approach from the standpoint of historical materialism, as this has been outlined above. He thus formulates the basic element of his approach:

> the socially necessary structures of thought of an epoch stand in the closest formal relationship with the forms of social synthesis of that epoch (1972, p. 20).

The social synthesis of capitalist society, he argues, is based on two forms of social organization: the exchange of commodities and the division of intellectual and manual labour. Hence the problem is to derive the categories of the modern natural sciences from those two forms of social organization.

Sohn-Rethel characterizes the structure of thought exhibited in modern natural science as 'mechanistic thought', and essentially accepts Kant's description of the forms of sensibility and of the categories of the understanding. Mechanistic thought, he observes, operates as an abstraction from the specific features of the multifarious different kinds of things, in order to reduce them to formal principles in terms of which they may be compared and related to one another. By such a process of abstraction, purely atomic concepts are arrived at which have only formal qualities, number, motion, and causality. Sohn-Rethel argues that this abstraction derives from and corresponds to the 'real abstraction' of commodity exchange. Following Marx's analysis of the commodity in the first chapter of *Capital*, he shows that the value of a commodity is an abstraction from all the various specific features of different products of human labour, which reduces them all to a common denominator in order that they may be compared for the purposes of exchange. The argument is followed through in an analysis of a number of different categories, which need not be recapitulated here. The conclusion is that the categories of mechanistic thought derive, not from a Kantian transcendental consciousness, nor from human productive labour in general, but from the specific form of social organization (*Vergesellschaftung*) of commodity-producing societies, namely exchange.

The need for exchange is derived from the division of labour, in which the economy consists of independent private producers producing for a market on which they exchange their products. However, for Marx and Engels

> division of labour only becomes truly such when a division of material and mental labour appears. From this moment onwards consciousness *can* really flatter itself that it is something other than consciousness of existing practice, that it *really* represents something without representing something real; from now on consciousness is in a position to emancipate itself from the world and to proceed to the formation of 'pure' theory (Marx and Engels, 1968, p. 43).

They meant by 'pure' theory such things as theology, philosophy, and ethics. Sohn-Rethel pushes the argument a stage further, and suggests that mechanistic thought, and the natural science based on it, is also a form of 'pure' theory which rests on the division of intellectual and manual labour. The natural sciences in their abstract form which they have taken on in commodity-producing societies are not derived from the immediate labour process, because the labour process takes a form which is under the control of those who take no direct part in it. Those who direct production and who develop the scientific technology which increases the productivity of labour work in the service not of labour but of capital. In this argument, Sohn-Rethel is building on some remarks of Marx:

> It is a result of the division of labour in manufacture that the worker is brought face to face with the intellectual potentialities of the material process of production as the property of another and as a power which rules over him. This process . . . is completed in large-scale industry, which makes science a potentiality for production which is distinct from labour and presses it into the service of capital (Marx, 1976, p. 482).

The argument is supposed to explain both the abstract categories of mechanistic thought and the contradictory relationship of natural science to productive labour: the fact that as 'pure' theory it is technically applicable to the labour process but only in the form of capital which is alien to the worker and subjects her or him to its control.

Sohn-Rethel's analysis, while partly systematic, in the sense of attempting to formulate homologies between the abstractions of mechanistic thought and of commodity-exchange, is also historical. In the second part of his book he presents historical material which is designed to show that mechanistic thought did indeed arise under conditions of early commodity-exchange, reaching its fully developed form in the seventeenth century.[7] Since mechanistic thought arose under those conditions, and remains systematically related to commodity-exchange and the division of intellectual and manual labour, one would expect Sohn-Rethel to take up Marcuse's theme and argue that the abolition of commodity production, exchange, and the division of intellectual and manual labour under socialist forms of social organization would also involve the development of a form of natural science not characterized by the abstraction of mechanistic thought. This indeed he does, but with such a degree of both understandable and yet startling vagueness that his speculations scarcely merit discussion (1972, pp. 207f.).

The general conclusion which may preliminarily be made on this first strand of thought concerning the critical epistemology of the natural sciences is that, while Marcuse, Sohn-Rethel, and before them Marx, were able to make plausible the historicization of synthetic activity and its relation to forms of productive labour and of the social organization of productive labour, they were much less successful in making plausible a conception of an alternative, non-alienated, or non-abstract natural science which an emancipated society would either require or would generate. With that preliminary conclusion we may pass over to the second strand of critical theory's thought concerning the epistemology of the natural sciences.

3.3.2. Habermas' Theory of the Technical Interest

Marcuse and Sohn-Rethel were concerned to show that the 'technical reason' or 'mechanistic thought' incorporated into what they took to be the existing form of the natural sciences are 'ideological', that they serve the particular interests of capital, the party, or the technocracy rather than those of the whole society. The implications of this approach tend necessarily towards speculation about the possibility of a 'non-ideological' form of the natural sciences, a form which, to take the example of Marcuse, would be guided by a desire for a 'liberating' rather than a 'repressive' mastery of nature. This speculation has not met with notable success.

In Habermas, whose work we may take as constitutive of the second strand of thought concerning the natural sciences stemming from the tradition of the critical theory of the Frankfurt School, the emphasis is quite different. While Habermas is quite prepared to note deleterious social consequences of the existing organization, control, and research orientations of the natural sciences (1971b, pp. 50ff.), he does not move beyond those essentially sociological, however critical, observations toward a consideration of forms of the natural sciences which might differ on either an epistemological or a methodological level. On the contrary, some of his ideas have been developed polemically against Marcuse's notions of an 'alternative' science or technology. In 'Technology and science as "ideology"', Habermas examines Marcuse's attempts to elaborate such notions and concludes that

> the idea of a New Science will not stand up to logical scrutiny any more than that of a New Technology, if indeed science is to retain the meaning of modern science inherently oriented to possible technical control (1971b, p. 88).

In fact, Habermas evinces no strong interest in more subtle distinctions within the philosophy of the natural sciences, and on a descriptive level he, like many other writers within this tradition, is prepared to accept the methodological formulations of writers who on more general epistemological grounds are attacked as 'positivist'. The concern is rather to work out fundamental differences between the natural and social sciences at an epistemological level, in order to demonstrate the inappropriateness of the methodology of the former in the context of the latter.

Habermas' general epistemological critique of positivism, like that of other writers within the tradition of critical theory, is essentially Kantian. The central fault of positivism is 'objectivism',

> an attitude that naively correlates theoretical propositions with matters of fact. This attitude presumes that the relations between empirical variables represented in theoretical propositions are self-existent (1972, p. 307).

Following Kant, and after him Husserl, Habermas criticizes objectivism for concealing or failing to treat the problem of the constitution of realms of facts, the problem of synthetic activity. The approach to this problem, however, can be neither the purely Kantian one, which ascribes synthetic activity to a transcendental consciousness as such, nor the purely Marxist one, which ascribes synthetic activity to the historically changing and developing forms of productive labour. The reason for the rejection of the Marxist approach, as Habermas himself reconstructs that approach, appears to be twofold: the first is that it seems to force Marx into comprehending human history as a continuation of natural history, thus treating cultural development as a mere corollary of a rising level of labour productivity, which involves "the reduction of the self-generative act of the human species to labour" (1972, pp. 41f.). The second is that it fails to distinguish between the 'life function' of labour and the 'cognitive function' of knowledge: it thus involves the

> mistaken notion that the synthetic functions of social labour must be analyzed on the same level as the process of social labour itself (Habermas, 1973a, p. 166).

The alternative to Kant and Marx proposed by Habermas is the self-reflection of the sciences of their own transcendental-logical conditions; that is, it is taken as given that the sciences, in the present context the natural sciences, have developed and have made successful progress, and the procedure is to examine the methodology, or the logic of inquiry, of the sciences with an eye to the question of what synthetic activity must have been achieved by what kind of subject in order that that success should have been possible.

Habermas implements this approach by an analysis of Peirce's study of the logic of scientific progress. Peirce identified three forms of inference that came into play in the course of scientific progress: deduction, induction, and abduction. Through deduction we apply existing theories to particular cases; no new knowledge is generated in this process. Through induction we check the agreement of hypotheses with the facts in order to evaluate them. Abduction is the key to scientific change, however, since it is conceived as the process of discovering new hypotheses, and hence of extending the content of theories about reality. Together these three forms of inference constitute a descriptive methodology for scientific research which, despite obvious differences as well as similarities, has the same epistemological status as the methodological rules propounded by Popper in the *Logic of Scientific Discovery*. The problem, however, is not to produce a descriptive methodology, but to give that methodology epistemological foundation.

In the case of Peirce, this involves an answer to the question of how it has been possible successfully to follow these methodological rules. This, as Habermas explains, is not an empirical question in the first instance, but a transcendental-logical one (1972, p. 116); he quotes Peirce:

> . . . What makes the facts usually to be, as inductive and hypothetic conclusions from true premises represent them to be? Facts of a certain kind are usually true when facts having certain (logical) relations to them are true; what is the cause of this? That is the question (Peirce, 1931–35, Vol. V, p. 341).

But whereas for Kant the answer to this kind of question is to be found in the necessary faculties of a transcendental consciousness, the answer Habermas derives from Peirce is a pragmatist one: the scientific method, and the concept of truth which characterizes propositions arrived at by the scientific method, is derivable

> only from the *objective life-context* in which the process of inquiry fulfils specifiable functions: the settlement of opinions, the elimination of uncertainties, and the acquisition of unproblematic beliefs – in short, the fixation of belief. The objective context in which the three modes of inference fulfil this task is the behavioural system of purpose-rational action (Habermas, 1972, pp. 119f.).

The search for beliefs which have the form of universal laws, which is what both Peirce and Habermas accept as an adequate account of scientific method, derives its meaning from the fact that such beliefs can function as practical beliefs; they can be acted upon, and the results of that action can be fed back into the search for improved beliefs, improved in the sense that they may be acted upon more reliably. And abduction only makes sense as a means of generating new hypotheses if nature itself is attributed with operation according to such 'rules' of practical action, which are then interpreted as causal laws; and to make this attribution is to constitute the natural environment as one that may be instrumentally manipulated. Causal laws are thus to be seen as recipes for producing effects, recipes which nature itself follows to produce its own effects and which people can discover to produce the effects they desire (1972, pp. 125ff.; cf. Gasking, 1955).

Pragmatism is thus interpreted in transcendental-logical terms; the behavioural system of instrumental action is the site of synthetic activity, and generates the categories which act as standards of validity for the empirical statements of the natural sciences. But these categories do not have transcendental necessity for any consciousness whatever, as in Kant; rather they have transcendental necessity only within the context of instrumental action. Unlike the case of Popper's 'one step from the amoeba to Einstein', however, pragmatism must interpret the relationship of scientific knowledge to instrumental action in contrast to the relationship of instinct to animal behaviour. In this way, Habermas introduces his theory of the 'technical interest' as follows:

> If we regard the function of knowledge in this way as a substitute for instinctive behavioural steering, then the rationality of feedback-controlled action is assessed with regard to the realization of an *interest* that can be neither a merely empirical nor a pure interest. If the cognitive process were *immediately* a life process, then the realization of this knowledge-constitutive interest would lead to the direct gratification of a need just as an instinctual movement does. The interest, however, when realized, leads not to happiness but to success. Success is assessed with regard to problem situations that have both a life function and a cognitive function. Thus 'interest' can be neither classed with those mechanisms of steering animal behaviour that we can call instincts nor entirely severed from the objective context of a life process. In this at first negatively delimited sense we speak of a *knowledge-constitutive interest in possible technical control*, which defines the course of the objectification of reality necessary within the transcendental framework of processes of inquiry (1972, pp. 134f.).

This is the technical interest which structures the transcendental framework of the natural sciences.

Since it is such a central component of Habermas' epistemology and will play an important role in the later argument, it will be useful to go back over the way in which Habermas generates the concept of a knowledge-constitutive interest and that of the technical interest in particular (cf. Stockman, 1978). It is fruitful to distinguish three strands in this derivation, which make it possible to pinpoint both the difference between Habermas' approach and that of Marcuse or Sohn-Rethel, and some difficulties with his approach which will occupy us later:

(a) Interests are defined on the level of the universal interests of the human species. In *Knowledge and Human Interests* we find: "I term *interests* the basic orientations rooted in specific fundamental conditions of the possible reproduction and self-constitution of the human species, namely *work* and *interaction*" (1972, p. 196). One set of such fundamental conditions relates to the relation of the human species to nature. Mankind exists within nature, but need not merely adapt to natural forces. "The human interests that have emerged in man's natural history . . . derive both from nature and from the cultural break with nature. Along with the tendency to realize natural drives they have incorporated the tendency towards release from the constraints of nature" (1972, p. 312). From this relationship is derived an interest in technical control, which "can be ascribed only to a subject that combines the empirical character of a species having emerged in natural history with the intelligible character of a community that constitutes the world from transcendental perspectives" (1972, p. 135). Technology as a 'project' realizing this interest can thus, in contrast to Marcuse, only be attributed to the human species as such, and not to any historically specific part of the species, such as a class. Similarly, science conceived of as producing knowledge from the transcendental perspectives of this interest must also be seen as a project of the species.

(b) Work, or instrumental action, which has already been identified as one of these basic orientations, is then shown to set the conditions in which one domain of possible objects of experience is constituted. In the 'Postscript' to *Knowledge and Human Interests*, Habermas summarizes this part of his argument as "the conjecture that the use of categories like 'bodies in motion' . . . implies an a priori relation to action to the extent that 'observable bodies' are simultaneously 'instrumentally manipulable bodies' . . ." (1973a, p. 174). Instrumental action upon this domain of objects is either successful or not, hence knowledge of this domain guided by an interest in the manipulation of objects can be corrected by feeding back information concerning success or failure into the processes generating knowledge. These processes are originally pre-scientific: they can be seen as a 'proto-physics'. In *Knowledge and Human Interests*, science is taken to be a straightforward continuation of the behavioural system of instrumental action, so that processes of inquiry are directly steered by the sucess or failure of technical manipulation. In the 'Postscript', however, Habermas recognizes that it is too simple merely "to conceive of the constitution of *scientific* object-domains as a continuation of the objectifications that are going on in practical life". For

when science seriously claims to be 'objective' it bases this claim on an (institutionally guaranteed) fundamental (and not only pragmatic) decision to suspend the pressures of experience and action, thereby enabling us to test *hypothetical* truth claims through discourses and to accumulate *corroborated* knowledge, or, which is the same thing, to build theories (1973a, p. 174).

In so doing, science constitutes an object-domain of theoretical entities, and the problem arises of how it is possible "to *interpret* the basic concepts of theories within the boundaries of a pre-scientifically attained objectification of experienceable happenings" (1973a, p. 175). An explicit connection is made between these first two strands of the argument: "The *universality* of cognitive interests implies that the constitution of object-domains is determined by conditions governing the reproduction of the species, i.e. by the socio-cultural form of life *as such*" (1973a, p. 177).

(c) The connections between knowledge in the natural sciences and the interest in possible technical control is established, as we have seen, by an examination of the relationship between the logical form of scientific theories and of the process of their refinement and the structure of instrumental action. The analysis, which, as we have seen, in *Knowledge and Human Interests* followed Peirce's pragmatist reformulation of the grounds of scientific method, had been more succinctly summarized in the earlier article 'Knowledge and human interests', which is worth quoting at length:

> In the empirical-analytic sciences the frame of reference that prejudges the meaning of possible statements establishes rules both for the construction of theories and for their critical testing. Theories comprise hypothetico-deductive connections of propositions, which permit the deduction of lawlike hypotheses with empirical content. The latter can be interpreted as statements about the covariance of observable events; given a set of initial conditions, they make predictions possible. Empirical-analytic knowledge is thus possible predictive knowledge. However, the *meaning* of such predictions, that is their technical exploitability, is established only by the rules according to which we apply theories to reality.
>
> In controlled observation, which often takes the form of an experiment, we generate initial conditions and measure the results of operations carried out under these conditions. Empiricism attempts to ground the objectivist illusion in observations expressed in basic statements. These observations are supposed to be reliable in providing immediate evidence without the admixture of subjectivity. In reality basic statements are not simple representations of facts in themselves, but express the success or failure of our operations. We can say that facts and the relations between them are apprehended descriptively. But this way of talking must not conceal that as such the facts relevant to the empirical sciences are first constituted through an a priori organization of our experience in the behavioural system of instrumental action.
>
> Taken together, these two factors, that is the logical structure of admissible systems of propositions and the type of conditions for corroboration suggest that theories of the empirical sciences disclose reality subject to the constitutive interest in the possible securing and expansion, through information, of feedback-monitored action. This is the cognitive interest in technical control over objectified processes (1972, pp. 308f.).

It is even more clear from this passage than in Habermas' discussion of Peirce that this derivation of the technical interest by a process of self-reflection of the natural sciences of their logic of inquiry depends for its force on an acceptance, at a descriptive level, of the account of the methodology, or the logic of science, provided by precisely those writers who are attacked as 'positivists'. Habermas does not question the account of the logical structure of theories provided by empiricism, but rather its objectivist illusion, its failure to reflect the underlying knowledge-constitutive interest.

At this point we may conclude our account of critical theory's positive contribution to the epistemology of the natural sciences. In their various ways, the critical theorists attempt to remedy the failure which they identify in their negative critique of positivism, a failure to go beyond descriptive methodology to provide a reflective theory, on the model of Kant's critical philosophy, of those elements of the knowing subject and of its relationship to its object, nature, which provide epistemological foundations for that methodology. All the writers whose work has been discussed, whatever their substantive differences, share this general philosophical approach. What they also share is an acceptance of the descriptive methodology of the natural sciences provided by those who are criticized as positivist. If we now turn to the critique of positivism mounted by scientific realism, we may point the way with the signpost, whose legend is to be elaborated, telling us that here we will find a philosophy of the natural sciences, one of the key characteristics of which is a reluctance to accept that descriptive methodology as adequate.

CHAPTER 4

THE ANTIPOSITIVISM OF SCIENTIFIC REALISM

4.1. Preliminaries on 'Realism'

With the term 'realism' complexities are encountered which are just as great as those with 'positivism' itself, if not greater. 'Realism' has undergone a historical development longer and more tortuous than 'positivism', and the variety of its contemporary use can only fully be understood by reconstructing that historical development. Without such understanding, a similar kind and degree of confusion can reign as in the case of 'positivism'. Thus Hilary Putnam observes:

> While it is undoubtedly a good thing that 'ism' words have gone out of fashion in philosophy, *some* 'ism' words seem remarkably resistant to being banned. One such word is 'realism'. More and more philosophers are talking about realism these days; but very little is said about what realism is (1978, p. 18).

One rhetorical difference in comparison with 'positivism' is that 'realism' is not in fashion merely, or even predominantly, as a term of abuse; as in everyday life – 'let's be realistic' – so in philosophy realism appears as a virtue to be claimed from many sides, though by no means from all. But given confusion over 'what realism is', what is claimed as virtuous need be not at all the same thing in all cases. Of the three traditions under discussion in the present work, two, scientific realism and critical rationalism, claim adherence to 'realism' in some form; to appreciate their differences in general will hence involve an understanding of their different interpretations of realism in particular.

However, it cannot be the intention here to do more than scratch the surface of the complexity beneath; while a rough outline of the main channels of historical development may be sketched out in an essay such as Raymond Williams' entry under 'Realism' in *Keywords* (1976, pp. 216ff.), a more thorough investigation would require and deserve considerably longer treatment than is possible here. It is nevertheless necessary to provide a very abbreviated initial map of those uses of the term which will enter into the discussion in this section and later; to this end *three* fields of use may be distinguished, which by no means exhaust the range of even contemporary usage, but which provide a starting-point for situating and identifying the kind of realism presently at issue:

(a) The first field of use for 'realism' is in the theory of universals (Woozley, 1967). It is one possible position to be taken on the problem of the nature of that generality which appears to be a necessary feature of experience and language. Our language consists primarily of general words, which seem to refer to abstract objects, qualities, properties and relations. Realism, in the sense in which it was used mainly in medieval philosophy, is the doctrine that the abstract objects indicated by such words have an objective existence independent of the knowledge we have of them or of the language in which we capture them. The predominant versions of realism in this sense were originally propounded by Plato and Aristotle. Plato's version was the theory of Forms, according to which each universal is a single substance or Form, which exists timelessly and independently of any of its particular manifestations. Aristotle's version denies universals such independent existence; they exist only as common elements of particular things. But they are nonetheless independent of our knowledge and language, and Aristotle conceived them as the objects of definitions that describe the essential natures of different kinds of things. The opposition to realism as a theory of universals comes from doctrines which stress that distinctions between different kinds of things are not independent features of the world but are rather made by people for their own purposes. Such doctrines comprise 'conceptualism', associated with the British empiricists, Locke, Berkeley, and Hume, which treats universals as general ideas formed by the mind, to which general words correspond, and 'nominalism', which carries the process further and argues that only words are general, and their meaningfulness can be accounted for without having to postulate mental ideas. It is usually accepted that the realist theories of universals, either in their Platonic or Aristotelian forms, were an obstruction to the development of modern empirical science, which was marked metaphysically by a shift to conceptualism or nominalism (Burtt, 1925; Koyré, 1968; Dijksterhuis, 1961). Hence it is widely thought that the debate over realism in this first sense is of purely historical interest.

(b) The second field of use for 'realism' is that concerned with questions of the existence of material objects and of our knowledge of 'the external world'. Realism in this sense is some version of the doctrine that material objects exist externally to human minds and independently of human sense experience. The contrast to realism is thus 'idealism', some version of the doctrine that "no such material objects or external realities exist apart from our knowledge or consciousness of them, the whole universe thus being dependent on the mind or in some sense mental" (Hirst, 1967, p. 77). The argument between realism and idealism represents precisely the kind of philosophical controversy which the logical positivists wished to reject as metaphysical, and was often used by them as an example of a "pseudo-problem" (Carnap, 1959; Schlick, 1959). When the logical positivists turned later to the question of what it was that protocol sentences, the basic sentences to which all scientific knowledge was supposed to be reducible, described, they were forced back to such 'metaphysical' problems, and produced the same range of theories concerning the nature of the objects of human perception as other philosophers not driven by positivist purity. Idealism was rather out of fashion, except in the form of phenomenalism, but the realist still faced a variety of problems, to which doctrines of direct realism, critical realism, and representative realism were the main responses (Halfpenny, 1976, pp. 57ff.; Hirst, 1967).

(c) The third field of use for 'realism' is located in debates over the cognitive status of scientific theories and the empirical status of theoretical terms in such theories (Halfpenny, 1976, pp. 149ff.; Nagel, 1961, pp. 106ff.). According to the deductive-nomological reconstruction of scientific explanation, explanation proceeds by the deduction of an event to be explained from a set of scientific laws in conjunction with a set of initial conditions obtaining before the event in question. At low levels of generality, the laws entering into such explanations may be treated as mere summaries of the covariances between observable events which have been discovered to hold good. It is characteristic of modern natural science, however, that higher levels of laws are proposed from which lower level laws may themselves be deduced and thus explained. Such higher level laws, or theoretical laws or theories, contain reference to apparently unobservable entities by means of what are referred to as 'theoretical concepts'. The problem arises concerning the status that should be accorded such theoretical entities and the laws in which they figure, a problem particularly acute for empiricists who wish to ground all scientific knowledge in a basis of sense experience. Radical empiricism would in principle consign reference to such entities to the category of metaphysics, or in more moderate versions attempt to show that the terms referring to them were in principle eliminable from scientific explanations. When this project was attempted, and judged to have failed, by the logical positivists, the problem of interpreting theoretical terms generated three main approaches. The first, which has been called 'descriptivism', attempted to show that theoretical language could be tied down to sense experience by empirical interpretations for some of the theoretical laws of a scientific theory. Scientific theories were conceived of as axiomatic systems, some of the axioms or theorems of which were tied to observational statements by correspondence rules or co-ordinative definitions. 'Realism' was taken over as the term to refer to the second approach, which interpreted theoretical terms as references to hypothetical entities which, although they happened at the moment to be unobservable, were really existing entities and observable in principle, if only experimental instruments could be developed to identify them. The link between theoretical entities and observational statements was thus not a definitional one, as with descriptivism, but might for example be interpreted as causal. The third approach, termed 'instrumentalism', differed from either descriptivism or realism in denying that theoretical laws had to be tied down to experience in order that their truth or falsity could be ascertained; theories have, on this view, no empirical content, but are merely instruments or computational devices for generating testable predictions (Nagel, 1961, pp. 129f.). Instrumentalism clearly has close affinities with what was earlier referred to as 'thesis-conventionalism', for it stresses that theories should be accepted or rejected not according to their truth or falsity but according to their usefulness.

This brief map of the usage of the term 'realism' is necessarily oversimplified and is intended as no more than a starting-point. Nor are the three fields of use which have been distinguished sharply demarcated from each other; in particular the third may be interpreted as a special case of the second, and just as the logical positivists treated the debate between realism and idealism as a typical example of a fruitless metaphysical argument, so it is possible to find echoes of this judgment in debates between realism and instrumentalism. Thus Nagel suggests that proponents of these views

> frequently disagree neither on matters falling into the province of experimental inquiry

nor on points of formal logic nor on the facts of scientific procedure. What often divides them are, in part, loyalties to different intellectual traditions, in part inarbitrable preferences concerning the appropriate way of accommodating our language to the generally admitted facts (1961, p. 141)

and concludes:

the opposition between these two views is a conflict over preferred modes of speech (1961, p. 152).

On the other hand, the third field of use for 'realism' is clearly limited to the interpretation of the theoretical laws of modern empirical science, and if it is the case, as is usually believed, that realism as a theory of universals is inimical to the development of modern empirical science, then realism in the third sense will have no relationship to realism in the first sense at all. If at the heart of modern empirical science lies an atomist ontology, then a realist theory of universals must be incompatible with empirical science, which will require instead a conceptualist or nominalist theory of universals as its metaphysical base.

What made it possible for logical positivists to argue that the choice between realism and idealism made no difference that *made* a difference was a consensus concerning the principle of verifiability and its interpretation. When this broke down, the 'metaphysical' problems re-emerged. Similarly, as Nagel makes clear, what makes it possible for him to argue that the choice between realism and instrumentalism makes no difference that *makes* a difference is a consensus concerning the methodology and logic of the natural sciences. This consensus comprises, minimally, a conception of scientific theories as deductively related systems of propositions, a deductive-nomological account of scientific explanation, and a view that science progresses by the use of the hypothetico-deductive method of experimental testing. Even if realists or instrumentalists do not agree with Nagel that they conflict merely over preferred modes of speech, they do tend to situate themselves within that consensus.

In recent years, however, this consensus has been broken by a theory of science which has taken for itself the name of 'realism'. In the light of what has been said so far, the appropriation of this name is thoroughly confusing, since so many realisms are already in the field. Nevertheless, there is no other way of referring to this theory, and indeed we will see that it draws on all three versions of realism that have previously been distinguished. In what follows, the term 'realism' will therefore be used to refer to this more recent theory, unless otherwise stated.

For some time, the philosophy of science, in concentrating on the logical structure of scientific theories and scientific explanation, has encountered considerable difficulties in accounting for a feature of scientific thinking which appears to be highly prevalent: the construction of models and the use of analogies and metaphors. These have seemed to be non-logical aspects of scientific thinking, and from the standpoint of logical positivism, logical empiricism, and associated 'logicistic' theories of science may be deemed irrelevant to the true business of scientific advance – heuristics of psychological interest, but no more. Realism, in its new form, accords models and analogies greater significance. "The Copernican revolution in the philosophy of science", as Rom Harré somewhat grandiloquently is wont to characterize the cause he champions

> consists in bringing models into the central position as instruments of thought, and relegating deductively organized structures of propositions to a heuristic role only, and resurrecting the notion of the generation of one event or state of affairs by another. On this view theory construction becomes essentially the building up of ideas of hypothetical mechanisms (1970, p. 116).

This is an explicit challenge to the consensus in the philosophy of science outlined above, a consensus which realists tend to refer to as 'deductivism', or, more loosely, 'positivism'. It is this antipositivism which it is the task of this section to explore.

It is worth making an initial remark on the choice of this realist theory of science as an object of discussion, which anticipates the examination of antipositivism in the social sciences in the following part of this work. Although realism in this form arose from dissatisfaction with the state of the philosophy of the *natural* sciences, and does appear to mark a significant development within that context, this cannot be its main interest for a sociologist. More importantly, there are signs, which will be discussed in the next part, that realism is increasingly being embraced by sociologists and possibly other social scientists as the source for a revitalized theory of the unity of science. This tendency is encouraged by the participation of some of the contributors to the realist theory of the natural sciences, who have extended their interest to the social sciences also. 'Unitarians', under pressure from their adversaries and criticized as 'positivists' (this time in the sense of advocates of the unity of science), have retaliated with a new, improved theory of science which may be thought to evade many of the criticisms brought by those adversaries. The impetus behind the present choice of realism as a form of antipositivism is thus not primarily its significance as a theory of the natural sciences, but rather a desire to consider the validity of this claim for its potential as the basis of the unity of science. This desire cannot, however, be implemented without a prior examination of realism's antipositivism in the philosophy of the natural sciences, to which we now turn.

4.2. Realism's Identification of Positivism

The target for realism's criticism, as has been mentioned already, is that complex of ideas which is seen to constitute a consensus in the philosophy of science. There is a degree of fluctuation in the terms used to characterize this consensus, which, while perhaps not important in itself, must be mentioned in the present context. It is referred to either as 'deductivism' or as 'positivism' (Harré, 1970, Chap. 1; Harré and Secord, 1972, Chap. 4; Bhaskar, 1978a, Introduction). Thus sometimes 'positivism' is used as equivalent to 'deductivism' as an all-embracing term for (most of) modern philosophy of science, while on other occasions it is used to refer to a more specific variant within the consensus, such as logical positivism or its later derivatives. In the latter cases, 'positivism' may be contrasted with other versions of 'deductivism', such as 'conventionalism' (Bhaskar, 1978a, p. 128; Keat and Urry, 1975, Chaps. 1, 3). There does not appear to be any coherent theory behind this usage, such as there is in Schnädelbach's account of conventionalism as the other side of the positivist coin, and hence itself a form of positivism. Rather, there may be a feeling that positivism in a narrower sense, such as logical positivism, exemplifies the features of deductivism most clearly, but that other forms of deductivism are not really significantly different for it to be worth making strict terminological distinctions. For the purpose of the present exposition of realism, this fluctuation of usage will be copied, and 'deductivism' and 'positivism' used interchangeably, except where it is otherwise indicated. We are thus dealing with a widened concept of positivism, which, as with critical theory's concept, includes many lines of thought whose authors would not be happy with that appellation.

According to Harré, deductivism can be expressed in terms of three basic metaphysical and epistemological assumptions which underlie modern philosophy of science, and which Harré believes are 'myths' (1970, pp. 5ff.). The *first* of these assumptions is atomism, a doctrine that the objects of experience are atomic impressions which, if they are impressions *of* anything, are impressions of atomic, independent events which are the ultimate particular elements of the world. To the extent that these atomic impressions, or simple ideas, are organized into generalizations, the general ideas in which such knowledge is couched do not describe general or abstract objects in the world, but can only be understood along conceptualist or nominalist lines. Such general knowledge can only be knowledge of regularities among atomic events; if we speak of laws of nature, these can only be conceived of as descriptions of the regular succession of kinds of events. The assumption of atomism automatically generates a problem of induction, to which some solution has to be found, and a problem of causation, to which the consistent response is a Humean theory of constant conjunction.

The *second* assumption is that propositions, expressed by sentences, are the only proper vehicles for rational thought. If sentences are to express knowledge, the terms they contain must be subject to the restriction that they can be tied down to atomic sense impressions, a restriction which in its purest form is a demand for ostensive definition. Such sentences can only be used in very limited ways:

> only those sentences were admitted which could be used to make statements the question of whose truth or falsity was in principle decidable by an empirical investigation of a very simple kind. If grammar is confined to the logic of true or false statements, it is evident that the only way sentences of the admitted kind can be arranged is in deductive systems; and the only relations that can hold between statements are those that derive from the need for statements to be consistent with one another, and the prohibition of inconsistency (Harré, 1970, p. 8).

The *third* assumption is

> the belief that the ideal form for knowledge, and particularly for scientific knowledge, is the deductive system (Harré, 1970, p. 8).

The belief is said to be most explicit in the work of Descartes, and to be associated with three closely related beliefs: that logical order somehow reflects natural order, that the method of science is that of analysis and synthesis, and that mathematics, which is deductively systematizable, constitutes the ideal of knowledge, towards which all other knowledge should develop.

From these three assumptions it is possible to derive all the central tenets of the empiricist analysis of the logic of scientific knowledge, for example the deductive-nomological theory of scientific explanation, according to which scientific explanation proceeds by the deduction of the explanandum from general laws and particular (for example, initial) conditions. In addition, deductivism includes a specific conception of the relationship between laws and evidence, which is usually expressed by realists as 'Nicod's criterion'; in Bhaskar's formulation this may be formulated as the two principles

> P_1: the principle of *empirical-invariance*, viz that laws are or depend upon empirical regularities; and
>
> P_2: the principle of *instance-confirmation* (or falsification), viz that laws are confirmed (or falsified) by their instances (Bhaskar, 1978a, p. 127).

Deductivism implies also such characteristic theses as the symmetry of explanation and prediction, and demarcation criteria, such as falsifiability, between science and non-science (Bhaskar, 1978a, pp. 127ff.).

These assumptions and theses taken as identifying characteristics of deductivism are clearly couched at a high level of generality, such that they might be thought to do injustice to specific schools of thought within the modern philosophy of science. In comparison with the conceptions of positivism advanced by either critical rationalism or critical theory, however, realism identifies positivism at what might be thought of as a lower level of philosophical argumentation. Both critical rationalism and critical theory claim to identify positivism as a (mistaken) account of what a theory of scientific knowledge should be like, although, as we have seen, critical theory denies the validity of critical rationalism's claim to have fully transcended the positivist account. Realism, on the other hand, appears to identify positivism as a set of substantive methodological theses, although admittedly somewhat general ones. This judgement is supported by a remark of one proponent of realism in both the natural and the social sciences, Russell Keat, who proposed to treat realism as a potential basis for a non-positivistic form of naturalism, where 'naturalism' was used to refer to the claim that 'the methodology of the natural sciences *can* . . . be applied to the social sciences'. By 'methodology', Keat meant

> neither very general questions about 'science as a rational enterprise', nor highly specific ones about research design and experimental techniques, but a 'middle range' of questions concerning the structure of scientific theories, their relationship to 'evidence', the nature of explanation, and so on (Keat, 1971, p. 3).

This aspect of realism is also reflected in the nature of its criticism of positivism.

4.3. Realism's Critique of Positivism

As Harré has pointed out, in developing its critique of positivism realism tends to adopt a method of 'confrontation' (1970, p. 5). Realist theories are placed side by side with positivist ones, and the realist theories are supposed to stand up better to the challenge of comparison. It is thus difficult to expound the negative critique of positivism in complete isolation from realism's own substantive theories. Nevertheless, two principles of criticism are clearly expressed in realist writings, and this section will provide an outline of these, leaving an account of a realist theory of science to the next section. The two principles are, firstly, that the assumptions of positivism generate problems that are insoluble in its own terms, and, secondly, that the assumptions of positivism generate descriptions of scientific activity and scientific reasoning which are simply incorrect, or at the very least partial and one-sided.

The problems generated by the assumptions of deductivism which realism claims are insoluble within the terms of those assumptions are problems familiar to all empiricist philosophy of science and have for years constituted the bread and butter of the subject. They all centre around the problems of induction and causation. The problem of induction arises directly from the assumption of atomism, and would not arise but for that assumption. The combination of an ontology of atomistic events and an epistemology of atomistic sensations or impressions makes it very difficult to account for the validity of any knowledge of a general nature, but deductivism requires such knowledge for its theory of explanation. It appears to be quite illegitimate to generalize from the knowledge of a set of particular events in the past to a universal statement which can provide knowledge of events in the future. Realism argues that as it stands this problem is insoluble. The main positivist attempts to justify induction, it is said, rest on the illegitimate assumption of precisely what it is necessary to prove: that the course of nature will not change and that the future will be like the past. Popper's attempt to solve the problem of induction, but without abandoning the deductivist assumptions, is likewise rejected, on two main grounds: firstly, that it does not escape the problem of induction anyway, since the assumption that a falsified theory will stay falsified in future is itself an inductionist or verificationist one (Bhaskar, 1978a, pp. 217ff.), an argument which Popper himself appears, reluctantly but consistently, to accept (1969, p. 248 n. 31; cf. Wellmer, 1967, pp. 200ff.); secondly, that it results in the claim that the only really significant evidence in science is that which would falsify a universal proposition (Harré, 1970, p. 24; 1972, pp. 48ff.). The alternative to this second conclusion is to allow confirmatory evidence to be significant, but that, it is argued, simply generates more problems such as Hempel's paradox, according to which there appears to be no more reason why the observation of a black raven should confirm the proposition that all ravens are black than the observation of a red herring or a white shoe (Hempel, 1965; Bhaskar, 1978a, pp. 222ff.), or Goodman's paradox, according to which there is no reason to believe that future examples of things that have hitherto turned out to be green, such as emeralds, will not after a certain date turn out to be blue (Goodman, 1954, pp. 73ff.; Harré, 1970, pp. 25ff.; Bhaskar, 1978a, pp. 222ff.). These problems, it is argued, arise from the atomistic assumptions and Nicod's criterion, and cannot be satisfactorily resolved without abandoning these.

Similarly, the assumptions of deductivism make it difficult to provide a theory of causation or to distinguish a statement of an accidental succession of events from a causal law. Starting from the Humean analysis of causation, empiricists have recognized the need to distinguish between those regular successions of kinds of events which justify the scientist in believing in a causal law and those that are accidental concomitances and do not justify such a belief; they have advanced a number of possible criteria on the basis of which this distinction could be made (Hempel, 1966, pp. 54ff.; Nagel, 1961, Chap. 4). Realists claim that within the assumptions of positivism no such criterion can satisfactorily solve this problem (Harré, 1970, pp. 6f.; 1972, pp. 115ff.).

Critical argument of this kind is of necessity inconclusive. Nothing in the way of proof that these problems are insoluble within the assumptions of deductivism is offered; indeed, it is difficult to know what the nature of such proof would be. The problems are familiar to all students of empiricist philosophy of science, as are the solutions, refutations, and counter-arguments advanced by those who have grappled with them. Hence only the very abbreviated account of this aspect of the realist critique has been thought necessary; it cannot be expected that any reasonably tenacious empiricist would be convinced by this form of argument, however detailed the examination of the problems arising from induction and causation might be. The fact that such problems have not been satisfactorily solved will not be interpreted by all concerned to imply that they cannot be solved. This sort of negative critique is insufficient, and Harré is right to argue that the way to make progress is to show that realist assumptions throw such problems into a quite different light.

The second general principle of the realist criticism of positivism is that its assumptions lead its adherents to descriptions of scientific practice which are incorrect. According to Harré, this principle is to be interpreted in a way which accords authority in providing a description of scientific practice to scientists themselves; he invites us to

> remember too that my aim is to preserve the persistent intuitions of scientists, as far as possible, against the philosophers. If received logic runs counter to an important intuition then I take it as a *prima facie* hypothesis that received logic is inadequate in some way (1970, p. 116).

However, this is a rather misleading way of stating the principle involved. Neither Harré nor other realist writers are in the habit of citing direct evidence concerning the 'intuitions' of scientists on the subject of their own scientific practice; nor, if Medawar's own expressed intuition is anything to go by, would they be likely to find much direct evidence, for

> if the purpose of scientific methodology is to prescribe or expound a system of enquiry or even a code of practice for scientific behaviour, then scientists seem to be able to get on very well without it. Most scientists receive no tuition in scientific method, but those who have been instructed perform no better as scientists than those who have not . . . The scientist is not in fact conscious of acting out a method (1969, p. 6).

Instead of conducting empirical investigation into scientists' expressed beliefs concerning scientific practice, as Harré's statement would seem to warrant, realists therefore follow Medawar's own suggestion, culled from Einstein, not to listen to what scientists say about scientific practice, but to look at what they do (Medawar, 1969, p. 10).

A glance at the way this suggestion is implemented by realists, especially in Harré's own writings, immediately reveals one facet of realist philosophy of science: realists tend to refer to and draw upon a much wider range of scientific research than is common in the 'positivist' writing they criticize. Logical positivists were both impressed and stimulated by the great advances in theoretical physics in the early part of this century, and the emphasis on physics as the ideal science has remained with all those developments in the logic of science whose ancestry can be traced back to logical positivism. The ideal of the reduction of all other sciences to physics has also been a feature of this philosophy, spurred on by the paradigm case of the reduction of thermodynamics to statistical mechanics. Realists, in contrast, while not abandoning physics, do tend to draw on a variety of other natural sciences as well, in particular the biological sciences, such as biochemistry, molecular biology, genetics, and anatomy. From this perspective, realists are able to argue that the assumptions of deductivism, themselves components of a specific philosophical tradition, have been reinforced through a partial concentration on certain aspects of physics, but give misleading accounts of scientific practice and of the structure of scientific theory and explanation not only in other branches of the natural sciences, but also of aspects of physics itself.

It is impossible to dwell here on the variety of examples which realists adduce to support the critique of positivism. The bare conclusion is that none of the three assumptions of deductivism are supported by the evidence of scientific practice (Harré, 1970, pp. 10ff.). Scientists do not work on the atomistic assumption that the only knowledge can be that of the regular succession of events, but instead seek knowledge of various kinds of relatively persistent and stable patterns and structures, the more permanent of which have constitutions which can be grasped as 'real essences', the different varieties of which are natural kinds. Secondly, the vehicles of thought with which scientists think are not only propositions, expressed by sentences, but rather complexes of propositions and pictures; the pictures are models of the structures of which the world consists, and the sentences express the possibilities of change which these structures are liable to or capable of. Thirdly, it is unusual to find in science these propositions expressed in the form of deductive systems:

In fact, in actual science, deductive systems are quite rare: fragments of such systems can be found in physics, but most scientists come up with descriptions of structures, attributions of powers and laws of change, related by having a common object, not by being then and there deducible from a common set of axioms (Harré, 1970, p. 10).

Thus deductivism generates an incorrect description of scientific practice; but what is more, in conjunction with the insolubility of the philosophical problems of positivism, it sets up a false ideal for scientific progress:

An ideal for scientific knowledge which effectively makes scientific knowledge unachievable is grotesque. It is the purpose of this book to offer an ideal for scientific knowledge which is derived from science, and not from mathematics, and which is not such as to make the achievement of scientific knowledge in its image impossible (Harré, 1970, p. 29).

On these grounds deductivism, or positivism, must be abandoned and realism embraced as the correct theory of science.

It is necessary at this stage to present a brief account of this realist theory of science. This will be done in the next section, before proceeding to an assessment of the three forms of antipositivism, each in the light of the other two.

4.4. Beyond Positivism: The Realist Theory of Science

On deductivist assumptions, scientific explanation is a purely logical procedure of the deduction of an explanandum from a set of universal laws and initial conditions. To explain an event, according to this schema, is equivalent to having had sufficient grounds for expecting it to have happened; thus, with a simple change of tense, an explanation of this form can be transformed into a prediction, a statement that there are sufficient grounds to expect an event to happen in the future. On the basis of evidence of the kind of reasoning that is actually taken to be a good explanation in various branches of the natural sciences, realists find this account of the nature of scientific explanation inadequate. Initially it is argued that deducibility of an explanandum is simply an insufficient condition for an argument to count as an explanation, partly because quite trivially valid examples of deductions from true premises can be imagined which no-one would want to call explanations, and partly because there exist non-trivial cases of startling and accurate predictability of natural events, such as was achieved by Babylonian astronomy, which again we would be unhappy to call explanatory. Scientists, according to realist argument, may sometimes seek explanations which have the form of deductive reasoning, but this is not a central component of their explanatory endeavours. Rather,

> scientists, in much of their theoretical activity, are trying to form a picture of the mechanisms of nature which are responsible for the phenomena we observe. The chief means by which this is done is by the making or imagining of models (Harré, 1970, pp. 34f.).

Models, as we have seen, are a feature of scientific thought which positivist philosophers of science have, to be sure, remarked upon, but which they have had difficulty in understanding. From a deductivist point of view they appear to have no function in the logical process of scientific explanation, and, in accordance with the positivist tendency to dichotomize scientific reasoning into a logical component, which can be the subject of analysis by philosophers of science, and a psychological component, which has no philosophical significance but which might be of interest to psychologists concerned with scientific investigation, such philosophers have had to allocate the construction and use of models to the psychological category. Thus models might be of heuristic value in helping scientists to generate theories or hypotheses which they can then test, or models might provide a refuge and an aid to familiarity for scientists dissatisfied with the cold logic of an axiomatic calculus. Pierre Duhem went so far as to see this psychological use of models as a fundamental point of difference between English physicists and their French and German counterparts, and to deprecate their use as an abandonment of clear logical thought (1954, pp. 70ff.). For realists, in contrast, models play an absolutely essential role in scientific thinking, since they are the vehicles for carrying pictures of the generative and productive mechanisms which scientists attempt to describe.

Scientists work on the assumption that the observable non-random patterns in nature, which they identify in the 'critical descriptive phase' of a scientific field (Harré and Secord, 1972, p. 70), are produced by the operation of such generative mechanisms. Scientific explanation of an event, according to this assumption, is not the deduction of a statement describing that event from a set of other statements, but a description of the mechanisms that produced it. In the case of some such mechanisms, their description is no more intrinsically problematic than the initial description of the non-random patterns of the events themselves, since the mechanism involved is itself open to observation in the same way as the events which it produces. Such mechanisms are 'accessible' to the senses, or to the senses when modestly extended by such instruments as microscopes or stethoscopes. More important scientifically are those mechanisms which are not directly observable in this way, but which are only indirectly observable, or 'quasi-accessible', or not observable at all, 'inaccessible'. In these cases, scientists are led to imagine what kind of mechanism might be responsible for the effects which they have identified; they do this, however, not in a completely unrestrained or uncontrolled fashion, but in the light of their knowledge of other kinds of mechanisms about which they already have some information. They operate with analogies of mechanisms with which they are already familiar, and they project such analogies onto the mechanisms which they are currently engaged in studying. They construct in this way models of the generative mechanisms which might be responsible for the effects they have observed, 'iconic models' or pictures of the structure and working of the mechanisms. A considerable amount of scientific research is then concerned with testing hypotheses which can be derived from such models, in particular existential hypotheses which propose the existence of specific kinds of entities or structures, and whose actual existence can be verified by experimental testing (Harré, 1970, p. 55 and Chap. 3).

The concept of a generative mechanism lies at the heart of the realist's attempt to make plausible, against the assumptions of deductivism, a conception of natural or physical necessity as the basis of the analysis of scientific laws. On deductivist assumptions, as we have seen, it has remained difficult to provide an account of scientific laws which distinguishes them from other kinds of generalizations. As Bhaskar puts it,

To ascribe a law one needs a theory, for it is only if it is backed by a theory, containing a model or conception of a putative causal or explanatory 'link', that a law can be distinguished from a purely accidental concomitance. The possibility of saying this clearly depends upon a non-reductionist conception of theory. Now at the core of theory is a conception or picture of a natural mechanism or structure at work. Under certain conditions some postulated mechanisms can come to be established as real. And it is in the working of such mechanisms that the objective basis of our ascriptions of natural necessity lies (1978a, p. 12; cf. Harré, 1970, pp. 101ff.).

The realist analysis of causality is thus a direct challenge to the regularity theory of causality stemming from Hume, which had thought, on atomistic assumptions, to have put paid to the notions of generation or production which underlie natural necessity. Realism claims to revive successfully these notions, and to make them the centre of the analysis of causal explanation made possible by the discovery of laws of nature.

In order to carry out this project, realism needs to put forward an alternative metaphysics to the ontology of atomistic events which, as was outlined above, is one of the main assumptions of positivism. This alternative metaphysics is one of powers and natures. The particular entities or structures which are postulated by scientists in the models of mechanisms underlying phenomena are conceptualized by realism to possess certain powers to act in given ways under given conditions; a simple example would be the theory that a particular kind of virus has the power to produce poliomyelitis. Powers, however, are not analyzed as dispositions on the procedure of Ryle, namely as conjunctions of hypothetical or conditional statements (Ryle, 1963, Chap. V); this analysis would remain within the assumptions of atomism. Rather powers ascriptions are analyzed in terms of the capacity of the material to which power is ascribed to produce certain effects or act in a certain way, "in virtue of its intrinsic nature" (Harré and Madden, 1975, p. 86), and in the appropriate conditions. It might appear, at first blush, as if this were precisely the kind of 'essentialist' pseudo-explanation characterized in such examples as 'opium produces sleep in virtue of its dormative properties'. Scientists evade this charge, according to realism, by treating a powers ascription not as a final explanation but as a stimulus for further research. If a virus of a certain kind is ascribed the power to produce poliomyelitis under appropriate conditions, this explanation for the disease demands to be elaborated by an examination of the 'intrinsic nature' of the virus in virtue of which it has that power. This will involve an investigation into the internal structure and composition of the virus; but since that internal structure and composition is itself a 'quasi-accessible mechanism', this investigation needs to proceed by the setting up of a model or picture of that internal structure which can itself be tested. The process of acquiring knowledge of the generative mechanisms which produce the observable patterns in the world of nature is therefore a stratified one; at any given stage knowledge might either be at the level of a model of a mechanism whose exact empirical status is in doubt, or at the level of the investigation of the internal structure of a mechanism whose existence has been established, but the precise working of which is as yet not adequately understood. The solution of problems at one level does not bring research to a halt, but sparks off a whole range of new questions at a deeper level. The stratification of scientific knowledge corresponds to the stratification of the natural world itself (Bhaskar, 1978a, pp. 163ff.). In Harré's earlier writings (1961) this was expressed as the doctrine of 'ontological depth'; a doctrine that there exist in nature domains of objects of different ontological depths, as opposed to the positivist doctrine that there exists only one such domain. In more recent formulations, the doctrine of the stratification of the world of nature is expounded in terms of the existence in nature of more or less stable structures, any one level of which can be both a component element of a larger structure and itself composed of an arrangement of smaller structures. The possibility exists, in science, of extending research in either direction on this ontological dimension; science can pursue either the 'regress of macroexplanation' or that of microexplanation (Harré, 1970, pp. 260ff.).[8]

The ascription of powers to a thing at any given ontological level provides a non-conventional criterion for distinguishing between those properties of a thing which are essential to it and those which are accidental (Bhaskar, 1978a, p. 88; cf. Brody, 1967). According to realism, science aims at the description of the real essences of kinds of things that exist in nature, where real essences are the intrinsic nature of kinds of things in virtue of which they have the powers to produce the effects which they are capable of producing. In contrast, the nominal essence of a kind of thing is conceived as those properties which are used in the empirical identification of it (Harré and Madden, 1975, pp. 101ff.). Real essences are not usually accessible to direct sense observation: they are discovered by an investigation, guided by the construction of models, of the 'fine structure' of the things which exhibit, under appropriate conditions, certain powers (Fisk, 1973, pp. 248ff.). Hence the discovery of real essences is the discovery of 'natural kinds'.

It is clear, in terms of my initial discrimination of different fields of application for the term 'realism', that this version of scientific realism is closely related to realism as a theory of universals. The distinctions made by scientists between different kinds of things are not, on this view, purely humanly made distinctions which may be more or less useful in the pursuance of particular purposes, as conceptualism or nominalism would have it, but refer rather to real distinctions in nature. However, neither the Platonic nor the Aristotelian analysis of universals and of how knowledge of them is attained is satisfactory for scientific realism, though it is closer to the Aristotelian version. Against Plato, but with Aristotle, realism argues that universals are the objects of a definition that describes a thing's essential nature (rather than an ideal Form that exists in a separate realm in its own right). Against Plato, but with Aristotle, realism argues that knowledge of universals is not prior to experience or innate, but is *a posteriori* and acquired gradually. However, for Aristotle, scientific knowledge consists, at least in part, in the classification of natural objects, on the basis of characteristics accessible to sense observation, into the real kinds, divided by genus and species, to which by nature they belong. For realism, such taxonomic knowledge is only a very early component of scientific knowledge, and is superseded by investigation into the internal structure of kinds of things. Realism's theory of universals is thus closer to Locke's account of real essences, although on conceptualist grounds this could not be taken by Locke to be a theory of universals. For Locke real essences were the internal constitutions of things, the arrangement of the minute corpuscles of which they were composed, which were responsible for the observable characteristics of things (1959, Vol. 2, pp. 26ff.). However, Locke did not conceive of real essences as explanations of active causal powers, nor did he believe that knowledge of real essences could be acquired. Realism takes these two further steps, and can therefore argue that science aims at the knowledge of the natures, or real essences, of things in virtue of which they have their powers. This knowledge is formulated as real definitions of natural kinds, definitions which are arrived at *a posteriori*. At any given moment, such knowledge is provisional.

> The real essence is only finally discovered when the analysis is complete. So far as we can tell, no analyses of things and substances have yet been completed, so our knowledge of the natures of things is as yet an approximation to knowledge of their real essences,

but can stand in for that knowledge in all relevant contexts (Harré and Madden, 1975, p. 102).

That is, real definitions are attempts to capture the real essences of kinds of things, but they are fallible attempts, and may require to be revised in the light of further information.

From the standpoint of this metaphysics of powers and real essences (or natures), causal laws cannot be interpreted as statements of the regular conjunction of events, as they are interpreted on deductivist assumptions. They must rather be analyzed as the tendencies of things in virtue of their real essences. In fact, Bhaskar argues that regular conjunctions of events are not just insufficient criteria for the existence of a causal law, they are also not necessary conditions. His argument depends on a distinction he makes between open and closed systems. Closed systems are systems in which constant conjunctions of events occur, open systems are those in which they do not. Most systems in nature are open; closed systems must normally be brought into existence by scientists by establishing experimental laboratory conditions. This enables Bhaskar to make an initial distinction between laws and conjunctions of events:

It is a condition of the intelligibility of experimental activity that in an experiment the experimenter is a causal agent of a sequence of events but not of the causal law which the sequence of events enables him to identify. This suggests that there is an *ontological* distinction between scientific laws and patterns of events (1978a, p. 12).

By experimentally isolating a particular generative mechanism, a scientist is able to test statements about the conditions under which the powers of the generative mechanism will be exercised and with what results; but a mechanism will only have determinate effects under such controlled conditions, where the operation of other mechanisms can be excluded or controlled. It follows that in open systems, such as normally pertain in the natural world, a large number of different generative mechanisms will be exercising their powers to cause effects at the same time; the actual patterns of events that result will be 'conjunctures', and cannot be treated as evidence for the nature of the generative mechanisms at work. Yet they are the same mechanisms as the ones isolated in closed systems:

It is only if we make the assumption of the real independence of such mechanisms from the events they generate that we are justified in assuming that they endure and go on acting in their normal way outside the experimentally closed conditions that enabled us to empirically identify them . . . Moreover it is only because it must be assumed, if experimental activity is to be rendered intelligible, that natural mechanisms endure and act outside the conditions that enable us to identify them that the applicability of known laws in open systems, i.e. in systems where no constant conjunctions of events prevail, can be sustained (Bhaskar, 1978a, p. 13).

The powers of things may exist in open systems without any specific outcome; and it follows that one may know that a law is effective without, in open systems, being able to make any prediction about an event.

The distinction between open and closed systems enables Bhaskar to state in alternative terms one of the central assumptions of positivism which realism opposes: the assumption of a closure (1978a, pp. 69ff., 127ff.). Positivism either explicitly, or more often implicitly, treats the world of nature as a closed system. Such a system, Bhaskar argues – against the explicit denial of such an assumption by, for example, Popper (1972, p. 215) – must be deterministic in the Laplacean sense, or 'regularity-deterministic'. A number of features of positivistic philosophy of science can be traced back to this assumption. That the laws derived from theories can be interpreted as empirical co-variances, for example, would according to Bhaskar only hold for closed systems; from which it follows that explanation is only deductive-nomological in closed systems. Another logically related consequence is that the thesis of symmetry between explanation and prediction only holds in closed systems, or, as Bhaskar puts it:

it is characteristic of open systems that two or more mechanisms, perhaps of radically different kinds, combine to produce effects; so that because we do not know *ex ante* which mechanisms will actually be at work (and perhaps have no knowledge of their modes of articulation) events are not deductively predictable (1978a, p. 119).

The deductive-nomological theory of explanation is thus a theory of closed systems, whereas Bhaskar places great emphasis on being able to

distinguish between two moments of the scientific enterprise (interpreted broadly): the moment of *theory*, in which closed systems are artificially established as a means of access to the enduring and continually active causal structures of the world; and the moment of its open-systemic *applications*, where the results of theory are used to explain, predict, construct and diagnose the phenomena of the world (1978a, p. 118).

A further consequence is that the relationship between explanation and falsification is also restricted to closed systems. Since laws are not interpreted as empirical co-variances (except under artificially experimentally closed conditions), they cannot be falsified by open-systemic instances; in fact 'all law-like statements whose antecedents are instantiated in open systems are false' (Bhaskar, 1978a, p. 133). It can be seen from these features of the assumption of a closure that this is another way of stating the assumption of positivism that scientific theories must be deductively related systems of propositions.

At the beginning of this section on realism, we quoted Putnam's remark concerning the fashionable usage of the term 'realism' and its divergent meanings. It might be an implication of this remark, if Putnam is right in his assessment of fashion, that realist theories of science are now widespread in the ranks of philosophers of science. This assessment is supported by Frederick Suppe, in his very comprehensive survey of developments in the philosophy of science. He identifies, in the years following the symposium reported in his book, a number of trends, of which the first three are:

> (1) Positivistic philosophy of science has gone into near total eclipse;
> (2) the more extreme *Weltanschauungen* views of Feyerabend, Hanson, and Kuhn no longer are serious contenders for becoming a replacement analysis; and
> (3) philosophy of science is coalescing around a new movement or approach which espouses a hard-nosed metaphysical and epistemological realism that focusses much of its attention on 'rationality in the growth of scientific knowledge' and proceeds by the examination of historical and contemporary examples of actual scientific practice (Suppe, 1977, p. 618).

This may well be the case, so long as the first trend is taken to be the eclipse of the tradition of logical positivism and logical empiricism and of the 'received view' of scientific theories. But this must not, in the present context, be taken to mean that the form of realism which has been outlined in a highly compressed manner in the last few pages has become the new dominant view in the philosophy of science. Suppe in fact relegates a discussion of Harré's approach to a brief foot-note, in which he suggests that "it has not been particularly influential in shaping what I take to be the most important recent developments and trends in philosophy of science" (1977, p. 629 n. 36). Certainly a very wide range of theories of science may be gathered under the umbrella of 'realism', not all of them accepting all or even most of the main tenets of realism in Harré's sense and that of other writers working approximately in the same framework. Even between such writers there are significant differences of emphasis and approach. Other writers diverge even more sharply from the version of realism outlined above. Mary Hesse, for example, while she shares the realist rejection of deductivism, substituting a network model of theories drawn from Duhem and Quine, will have no truck with the neo-Aristotelian theory of real essences; "theories about essences", she suggests, "are neither stable nor cumulative, and are therefore not part of the realistic aspects of science" (1974, p. 299). C. A. Hooker (1974), advancing a doctrine which he calls 'systematic realism', incorporates aspects of neo-Aristotelianism, but rejects on the other hand a theory of a hierarchy of levels of reality which bears a strong resemblance to the theory of ontological depth, or stratification of nature, which plays an important part in Harré's and in Bhaskar's versions of realism. One could go on elaborating versions on this realist theme, but this would serve no useful purpose here. As has already been explained, it was decided to outline the particular version of realism stemming from the work of Harré, not primarily because of whatever significance it may or may not have within the present state of the philosophy of the natural sciences, although certain tentative suggestions will be put forward on this below, but because, both through the writing of Harré and his associates and through a number of other channels, this version of realism has entered the debate within the philosophy of the social sciences, and has been advanced as the basis for an antipositivist theory of the unity of science. On the principle that theses concerning the relationship between the natural and the social sciences depend as much on the view taken of the former as on the view taken of the latter, it is essential to make an assessment of realism as a philosophy of natural science before proceeding further. This question will be taken up in the next section.

CHAPTER 5

DISCUSSION: ANTIPOSITIVISM IN THE PHILOSOPHY OF THE NATURAL SCIENCES

In the first three chapters of this part, three different philosophical approaches to the theory of the natural sciences, each of which understands itself to be opposed to 'positivism', have been presented, and something of their background in different philosophical traditions sketched out. The task of this final chapter is to bring these three approaches into relation to each other, to draw preliminary conclusions on antipositivism in the theory of the natural sciences which can be carried forward to the discussion in the next part of this work on antipositivism in the philosophy of the social sciences. An initial point of orientation in this task of relating these three antipositivist approaches to each other is gained by recalling a point made in the introduction: that at the present time, when 'positivism' has become a term of abuse, the charge of being positivistic may be levelled against a theory which itself, in turn, conceives of itself as antipositivist and may return the charge against the accuser. It is, in fact, the aim of this work to contribute to the clarification of the confusion that thereby results.

It is convenient to begin by examining the relationships that may be discerned between critical rationalism and scientific realism. The observation must immediately be made that critical rationalism, too, stresses its own commitment to realism. Popper's claim to be a 'metaphysical realist' was in fact, as has been noted, one component of his rejection of the term 'positivist' when applied to himself, for metaphysical realism, in the sense in which it may be opposed to idealism, was precisely the kind of metaphysical doctrine which the logical positivists condemned as meaningless. In relation to the forms of realism which have been the focus of Chapter 3 however, it is clear that Popper was using the term in ways which are, on the surface at least, quite compatible with the basic assumptions of deductivism. Popper's self-definition as a realist takes on elements of realism drawn from the second and third of the three fields of use of 'realism' distinguished at the beginning of that section; namely, realism as contrasted with idealism, a doctrine of the existence of a world independent of the human mind, of which human beings may yet gain knowledge; and realism as contrasted with instrumentalism, the doctrine that the terms referring to theoretical entities and structures in scientific theories actually do, or may, refer to the real furniture of the natural world, and that theories are not just computational devices for inferring between observational statements. Realism in these senses is of course not denied by the scientific realists whose theories have been outlined, but a commitment to it is insufficient to escape a charge of positivism, for of course critical rationalism by no means constitutes a departure from the basic assumptions of deductivism.

Popper developed his view of realism by contrasting it with two other views of knowledge which he termed 'essentialism' and 'instrumentalism'. Essentialism is characterized by three basic doctrines: "(1) the scientist aims at finding a true theory or description of the world (and especially of its regularities or 'laws'), which shall also be an explanation of the observed facts". This doctrine is accepted by Popper and also forms part of his realist view. "(2) The scientist can succeed in finally establishing the truth of such theories beyond all reasonable doubt". This doctrine is rejected on fallibilist grounds: there can be no final knowledge, all scientific knowledge remains conjectural and open to further test. "(3) The best, the truly scientific theories, describe the 'essences' or the 'essential nature' of things – the realities that lie behind the appearances. Such theories are neither in need nor susceptible of further explanation: they are ultimate explanations, and to find them is the ultimate aim of the scientist" (1969, pp. 103f.). This doctrine is rejected, not on the grounds that essences do not exist – this is neither affirmed nor denied, though the doctrine that there exist levels of reality behind the appearances which nonetheless may be discovered is an element of Popper's own realism – but on the grounds that belief in essences is not helpful to the scientist and more likely to be harmful, in that it prevents fruitful questions from being raised. A belief in essences as ultimate explanations prevents the scientist from asking questions about the causes of the behaviour of essences.

The realism advanced by Harré and others, which as we have seen does place emphasis on the discovery of essences, has therefore been criticized as 'essentialist' on Popperian terms (Miller, 1972). There are at least two responses that might be made to such criticism. Firstly, as Suppe has pointed out, Popper's critique of essentialism does not rule out the existence of real essences. But on the assumption that such essences do exist,

> it is hard to see how the truth of theories can be accounted for without recourse to essences: if the theory is true under such a position, it must be so because it correctly describes the essences – this being so even if it is granted that we can never determine whether a theory in fact does correctly describe the essences (Suppe, 1977, p. 168).

Secondly, in the form in which realism of the Harré type proposes the discovery of essences, essences do not constitute ultimate explanations. As we have seen above, the description of the essential nature of a thing in terms of its powers to cause effects leaves open the question of the internal structure of the thing in virtue of which it has those powers, and this oscillation of scientific investigation between powers and mechanisms is conceived by the realist to have, in principle, a theoretical limit so distant that it cannot act as a brake to scientific curiosity in the way that Popper supposes. We can conclude that Popper's critique of 'essentialism' is inadequate to dispose of this form of realism.

The second view of knowledge rejected by Popper is 'instrumentalism'. According to the instrumentalist view, scientific theories are merely instruments which enable the scientist to deduce phenomena from prior phenomena; a universal law or a theory is not a statement which is true, or which can be proved false, but rather 'a rule, or a set of instructions, for the derivation of singular statements from other singular statements' (1969, p. 108). Popper's reply to instrumentalism

> consists in showing that there are profound differences between 'pure' theories and technological computation rules, and that instrumentalism can give a perfect description of these rules but is quite unable to account for the difference between them and the theories (1969, p. 111).

Instrumentalism cannot account for the procedure, central to Popper's own account of scientific methodology, by which scientists put theories to the most rigorous test possible, even testing out the most remote implications of a theory; it is thus unable to account for the scientist's interest in truth and falsity, especially as a theory may continue to be used for the purposes of practical application even after it has been refuted, so long as it is used only within the limits of its applicability. Instrumentalism, unable to develop a concept of falsification, cannot account for scientific progress, since it cannot explain why scientists attempt to develop theories beyond the limits of application; it is thus a conservative doctrine, inducing 'complacency at the success of applications' (1969, p. 114).

In contrast, Popper advances his own, third view, his own version of realism. In doing so he accepts the first, realist, doctrine of essentialism, namely that scientists aim to find true theories of descriptions of the world. As he put it in *The Logic of Scientific Discovery,*

> Theories are nets cast to catch what we call 'the world': to rationalize, to explain, and to master it. We endeavour to make the mesh ever finer and finer (1968, p. 59).

Theories ascribe reality to the world described by them in the sense that, although they remain hypothetical and conjectural, they claim to be describing something real. The reality they claim to describe is, like that of the realists described in the previous chapter, stratified;

> the world of each of our theories may be explained, in its turn, by further worlds which are described by further theories – theories of a higher level of abstraction, of universality, and of testability ... Since according to our third view the new scientific theories are, like the old ones, genuine conjectures, they are genuine attempts to describe these further worlds. Thus we are led to take all these worlds, including our ordinary world, as equally real; or better, perhaps, as equally real aspects or layers of the real world (1969, p. 115).

In contrast to those realists, however, this conception of the stratification of nature does not form part of a rejection of deductivism, but rather remains at one with a realist interpretation of the theoretical entities referred to by theoretical terms incorporated into deductively related hierarchies of universal laws.

Apart from their shared elements of realism in more general terms, scientific realism and critical rationalism can be said to differ mainly as accounts of the methodology of the natural sciences. Critical rationalism conceives of scientific method as a process of conjecturing universal statements as hypothetical scientific laws which are then tested against their instances. Scientific realism conceives of scientific method as a process of moving towards real definitions of the real essences of natural kinds, by means of the conjecturing of generative mechanisms which produce phenomenal effects, and the testing of existential and descriptive hypotheses concerning the structures postulated in models of such mechanisms. But lying behind this difference in descriptive methodology there is also a philosophical difference of what is involved in putting forward a methodology. Scientific realism understands itself as advancing a methodology which is a rational reconstruction of the actual practice of scientists; if a philosophy of science cannot capture the main features of this practice, or if it results in prescriptions for practice which run counter to what scientists actually do, then it is *ipso facto* inadequate or misguided. Critical rationalism, on the other hand, conceives of its methodological rules as conventions, and conventions to be agreements that are concluded by decision, made by people who share certain purposes or ends for which the methodological rules are appropriate means. Methodology, from this point of view, goes beyond descriptive rational reconstruction to become normative; it is thus possible for a significant component of scientific practice to diverge in actuality from these normative rules without thereby affecting their validity. In *The Logic of Scientific Discovery*, Popper had dismissed a methodology which was based on a description of the actual behaviour of scientists as 'naturalistic', arguing that the fact that scientists do follow certain methodological rules does not make those rules rationally justified (1968, pp. 51ff.). Instead, critical rationalism was to derive the justification of its proposals for methodological conventions from a more general and philosophical theory of knowledge. As we have seen, critical theory on the one hand rejects critical rationalism's claim to have provided such a theory of knowledge, but on the other hand commends critical rationalism in comparison with logical positivism for attempting a process of transcendental reflection, even if this process is broken off part way. If it would be correct to argue that, from the standpoint of critical rationalism, scientific realism appears to be a naturalistic methodology, in Popper's terms, then it might be also correct to argue that, from the standpoint of critical theory, scientific realism is that much more positivistic than critical rationalism in virtue of its failure to move beyond rational reconstruction of descriptive methodology towards a critical philosophical theory of knowledge. This argument will now be elaborated, by means of an examination of realism from the point of view of critical theory.

In the dispute between critical rationalism and critical theory, the *Positivismusstreit*, the realist doctrine of the former played a significant role. As Schnädelbach has pointed out,

The contributions of Hans Albert in the *Positivismusstreit* make it clear that the realist thesis of the theory-independence of the object of knowledge provides the grounds for the denial of all technicist-instrumentalist interpretations of knowledge, among which

he and Popper number both pragmatism and, in the polemical context of that debate, also Habermas' theory of the technical cognitive interest (1972, p. 90).

Albert does not of course thereby deny that the knowledge generated by natural scientists *can* be given technical application; like Popper he would provide an account of technical application in terms of a transformation of the deductive-nomological schema of explanation, according to which the explanandum becomes a set of specifications for a required state of affairs, and the initial conditions are interpreted as those technical conditions which need to be realized in order, under the given physical laws, to bring about that desired state of affairs (Popper, 1972, pp. 352f.). But according to Albert successful technical action merely proves that the theories in which the physical laws are stated and which are relied upon by the technologist, are sufficiently good accounts of reality (1976b, p. 171). Scientific theories are increasingly adequate if they draw increasingly close to reality, in accordance with Popper's theory of verisimilitude (1972, pp. 47ff.); the fact that such theories can also be used for technical application is indeed a consequence of their logical form, but from the scientific point of view purely a by-product of scientific progress. To connect technical applicability intrinsically with the adequacy-criteria of a scientific theory is interpreted as a form of instrumentalism, and Albert relies essentially on Popper's critique of instrumentalism, as this was outlined above, in order to reject Habermas' theory of the technical interest (1976b, pp. 170ff.).

Popper and Albert believe it is necessary to cling to metaphysical realism as an antithesis to instrumentalism because of the association the latter doctrine has, at least in their scheme of things, with conventionalism, or more accurately with thesis-conventionalism. Thesis-conventionalism denies that there can be any independent test of a scientific theory, since the 'facts' that a theory has to be tested against have themselves to be theoretically interpreted, and it is always possible to maintain the internal consistency of networks of theoretical propositions by means of procedures of re-interpretation which Popper refers to as 'conventionalist strategems' and Albert as strategies of 'immunization'. In order to avoid the conservative implications that thesis-conventionalism shares with instrumentalism, the possibility of independent testing must be established. This is the task of metaphysical realism:

> The task of science, which, I have suggested, is to find satisfactory explanations, can hardly be understood if we are not realists. For a satisfactory explanation is one which is not *ad hoc*; and this idea – the *idea of independent evidence* – can hardly be understood without the idea of discovery, of progressing to deeper levels of explanation: without the idea that there is something for us to discover, and something to discuss critically (Popper, 1972, p. 203; cf. 1969, p. 177).

The idea of an independent reality appears then to be necessary to provide a realm against which hypotheses may be tested, and thus to underpin the fallibilism which is at the heart of Popper's theory of scientific progress.

This metaphysical theory of an independent reality, however, is systematically at odds, as Schnädelbach has argued, with Popper's own critique of positivism, according to which, as we have seen above, all observations of the facts are interpretations in the light of theories (Popper, 1968, p. 107 n.*3). Popper's criticism of logical positivists' account of protocol sentences, as basic statements which constituted the empirical basis of science, was precisely that there could be no pure independent givens, whether in sense experience or anywhere else, which could act as a theory-independent realm of control. Popper's own theory of basic statements is as conventionalist as the rest of his methodology; basic statements are those statements which are not in principle immune to test, but which a scientific community agrees to treat as sufficiently well documented to be withdrawn from test for the time being; but this is merely a conventionalist twist to the original positivist conception of the given, and structurally basic statements play the same part in the logic of science in both logical positivism and Popperian conventionalism. But if this is so, Schnädelbach continues, then Popper's metaphysical realism is methodologically meaningless, and indeed both Popper and Albert appear to accept that the methodological theory of critical rationalism can be stated without it (Popper, 1972, p. 203; Albert, 1976a, p. 234). It is difficult, following this argument, to avoid Nagel's conclusion (1961, p. 152) that the dispute between realism and instrumentalism is one over the preferred use of terms and nothing more.

The problem with metaphysical realism, from the standpoint of critical theory, is that it is a pre-critical philosophical position, since it attempts to conceptualize an independent reality without raising the question of the constitution of that 'reality' through synthetic activity. We have already followed through the argument of the critical theorists, that synthetic activity is inadequately conceived of as the mental activity of a consciousness in general, but, so far as the categories of the natural sciences go, should rather be conceived of as the technical, instrumental activity of members of the human species engaged in a productive relationship with nature. In his critique of realism, Schnädelbach develops this argument further. If we examine the conditions under which 'reality' is to be conceived as 'independent', we are led to reflect on the contexts in which experience of an independent reality may be gained. In this we may be guided by a remark of Popper himself:

> The disappointment of some of the expectations with which we once eagerly approached reality plays a most significant part in this procedure. It may be compared with the experience of a blind man who touches, or runs into, an obstacle, and so becomes aware of its existence. *It is through the falsifications of our suppositions that we actually get in touch with 'reality'*. It is the discovery and elimination of our errors which alone constitute that 'positive' experience which we gain from reality (1972, p. 360).

Although this is couched in the language of metaphysical realism, it can easily be translated into the pragmatic terms of instrumental action. The reality which the blind man becomes aware of is the reality constituted in the sphere of instrumental action, it is that reality which acts as a resistance to our attempts at technical control of material objects. It is indeed 'independent', not in a metaphysical sense, but in the sense that it constitutes a resistance to the implementation of any human will whatsoever in the behavioural context of instrumental action (Schnädelbach, 1972, pp. 97f.). Experience of the success and failure of instrumental action thus provides evidence of the nature of the 'reality' constituted by synthetic activity in that context.

However, despite this criticism of the pre-critical nature of Popper's metaphysical realism, Schnädelbach is not satisfied to reject realism entirely. Instead he recalls Peirce's account of the pragmatic conditions under which it is meaningful to call things 'real'. The terms 'real', 'actual', and 'independent' are meaningfully used, not directly to describe objects of experience, but at a meta-level to reflect on the nature of that experience. As Kant argued, 'reality' (or 'actuality', *Wirklichkeit*) is one of the categories of modality, which differ from other categories in that

> in determining an object, they do not in the least enlarge the concept to which they are attached as predicates. They only express the relation of the concept to the faculty of knowledge (1929, p. 239).

'Real' is thus a modal concept. Following the pragmatist reformulation of the critical philosophy, elements of which were outlined earlier, the context of application of the term 'real' is not that of instrumental action itself, as would appear from Popper's account of the experience of the blind person, but of the reflective understanding of success and failure of instrumental action. For Peirce, inquiry was the process of attempting to restabilize beliefs when the habitual action based on those beliefs became unsuccessful. Research is thus directed towards the 'fixation' of belief in situations when doubt has been cast on their adequacy in contexts of instrumental action, and scientific research is characterized by being directed towards beliefs which are fixed in the sense of being immune to changes in arbitrary personal opinion. What is meant by beliefs being 'true' is that they can be accepted by all members of a community of investigators, without coercion, in the last analysis: truth is thus a limit concept, a concept which regulates the process of inquiry. Finally, 'reality' is conceived of as the object of true belief:

> The real, then, is that which, sooner or later, information and reasoning would finally result in, and which is therefore independent of me and you (Peirce, 1931–35, Vol. V, p. 311).

Reality does thus exist independently of its being known, but merely to say that, as the critical rationalists do, says nothing. It is necessary to go on to consider the nature of the research context, in which reality is the object of the true beliefs towards which inquiry is, in the limit, directed. And in the case of the natural sciences, that research context is defined as concerned with the manipulation of objects in the behavioural system of instrumental action.

Schnädelbach is thus led, by his critique of critical rationalism's metaphysical realism and his replacement of it by Peirce's reconstruction of realism, to concede to critical rationalism a distinction between science and technology. He distinguishes between 'technical action',

> at the basis of which lies an unproblematized unity of 'belief' and 'habit' in Peirce's sense, which is actualized in order to attain a goal for action which has been defined in advance;

and 'experimental action',

which in the widest sense is distinguished by the fact that that unity of belief and habit is itself problematized and becomes the occasion for inquiry, as Peirce conceives it: i.e. the active attempt at restabilization by means of the scientific method (1972, p. 104).

In the terms in which Habermas takes over and incorporates this distinction, technical action occupies the sphere of direct life-praxis, while experimental action occupies that of discourses in which beliefs are tested free from the immediate constraints of direct application, in particular free from the need to suspend methodological doubt for the sake of technical applicability (Habermas, 1973a, pp. 174f.). But this concession to critical rationalism does not reinstate an objectivist conception of knowledge: both technical and experimental action in the natural sciences remain, according to both Schnädelbach and Habermas, bound to the context of instrumental action, and the theory of the technical interest underlying knowledge in the natural sciences remains intact (Schnädelbach, 1972, pp. 104f.). In fact, if the experience of an 'independent reality' in the course of technical action is the experience of an obstruction or resistance to that action, then realism as a philosophical theory can be interpreted as an attempt to reflect upon the problems that are thrown up by obstructions to technical plans, and in the context of that reflective theory experimental action can be understood as 'reflected technical action', i.e. as technical action shifted to the meta-level of reflection, in which, by means of experiment, solutions to the problems raised by the failure of technical action are to be investigated. 'Realism' is thus taken over from critical rationalism and incorporated into critical theory's analysis of the technical interest.

It is in the light of this discussion of the critical theorists' attitude to critical rationalism's metaphysical realism that we may now turn to the relationships between critical theory and the scientific realism outlined in Chapter 4. It has already been shown that what distinguishes this form of realism from any of the other theories of the natural sciences presently under consideration is that it presents a quite different view of the methodology of the natural sciences. It has been stressed that the antipositivism of critical theory, for example, does not amount to a rejection of deductivism as a reconstruction of natural science methodology. Scientific realism, in contrast, conceives of its opposition to positivism specifically as a rejection of the deductivism which is common ground between critical theory and critical rationalism. It is not, therefore, surprising, but rather entirely consistent, to find those advocates of scientific realism who have encountered critical theory's theory of the technical interest including it under the positivism to which they counterpose their realism; for, as we saw when examining the way in which Habermas derives the notion of the technical interest, that derivation itself depends upon a deductivist account of scientific explanation and of the structure of scientific theories. Russell Keat and John Urry, for example, identify as a major defect in Habermas' theory of the cognitive interests the fact that

> Habermas' account of the nature of the empirical-analytic sciences is essentially positivistic. This is revealed partly by the way he characterizes the objects of these sciences, in terms of what is open to detection and manipulation in controlled experiments; partly by the way he describes their aim as the discovery of nomological knowledge. And we argue in part 1 that realism, at least in the natural sciences, is preferable to both positivism and conventionalism. Thus we regard Habermas' analysis of the empirical-analytic sciences as failing to challenge the positivist understanding of scientific knowledge, its view of theories and explanations, and its ontological and epistemological restrictions (1975, p. 227).

Similarly, Anthony Giddens has argued that

> Habermas (and Apel) tend to operate with too simple a model of the natural sciences, which are described in a traditional – even positivistic – way. Habermas in fact rarely discusses the natural sciences directly, referring to them mainly in relation to the form of knowledge-claim or 'cognitive interest', in technical control, associated with them (but also associated with other disciplines) . . . 'Explanation' in the natural sciences assumes various forms just as in other spheres of enquiry' (1976, pp. 67f.),

and in a later essay discusses, among other non-positivistic accounts of the natural sciences, the realist network theory of Mary Hesse (1977, pp. 77f.). From the standpoint of the realist theory of science, critical theory's epistemology, at least in the shape of Habermas' theory of the technical interest, is just as positivistic as critical rationalism or logical positivism, and to be rejected as such.

Before proceeding to an assessment of this new charge of positivism and its implications, the relationship between critical theory and scientific realism may be also regarded in the other direction. In particular, how does scientific realism stand in relation to critical theory's critique of critical rationalism's metaphysical realism? It was suggested earlier in Part II that the differences between critical rationalism and scientific realism are of two kinds: first, a difference in their accounts of the methodology of the natural sciences, and secondly, a difference in their conceptions of the philosophical status of methodological rules, with critical rationalism stressing the normative and conventional status of those rules, and scientific realism stressing their status as a model designed to rationally reconstruct actual scientific practice. Alongside these two kinds of differences, however, lies a common commitment to metaphysical realism, and the nature of this level of their respective doctrines of realism is remarkably similar. Scientific realists, in agreement with critical rationalists, stress the 'independence' of the 'reality' which is the object of science. This is made absolutely clear by Bhaskar:

> knowledge is '*of*' things which are not produced by men at all: the specific gravity of mercury, the process of electrolysis, the mechanism of light propagation. None of these 'objects of knowledge' depend on human activity. If men ceased to exist sound would continue to travel and heavy bodies fall to the earth in exactly the same way, though ex hypothesi there would be no-one to know it. Let us call these, in an unavoidable technical neologism, the *intransitive objects of knowledge* ... We can easily imagine a world similar to ours, containing the same intransitive objects of scientific knowledge, but without any science to produce knowledge of them. In such a world, which has occurred and may come again, reality would be unspoken for and yet things would not cease to act and interact in all kinds of ways. In such a world the causal laws that science has now, as a matter of fact, discovered would presumably still prevail, and the kinds of things that science has identified endure ... In short, the intransitive objects of knowledge are in general invariant to our knowledge of them: they are the real things and structures, mechanisms and processes, events and possibilities of the world; and for the most part they are quite independent of us. They are not unknowable, because as a matter of fact quite a bit is known about them. (Remember they were introduced as objects of scientific knowledge.) But neither are they in any way dependent on our knowledge, let alone perception, of them. They are the intransitive, science-independent, objects of scientific discovery and investigation (1978a, pp. 21f.).

These intransitive objects of knowledge, if they are actually to be known, must of course be conceptualized; they must be grasped in the language of theories and models which are the social productions of human beings, and these social productions are passed on from one generation of scientists to another, undergoing change and development in the process. These theories, models, and so on are termed by Bhaskar 'transitive objects of knowledge'. According to Bhaskar, an adequate theory of science must be able to sustain and reconcile both of these aspects of science. But it is clear from the above quotation that the distinction between transitive and intransitive objects of knowledge has already incorporated within it a form of metaphysical realism which, with its emphasis on 'independence', no critical rationalist would have difficulty accepting.

If this is so, it would appear that Schnädelbach's critique of realism ought to be equally applicable to the realism of the scientific realists as to that of the critical rationalists. In both cases, the doctrine of the 'independence' of the objects of knowledge, rather than being generated from a transcendental reflection on the conditions of the meaningfulness of the term 'independent' is merely a restatement of the doctrine of metaphysical realism itself, and cannot be stated independently of it (Schnädelbach, 1972, p. 97). Or in other words, in neither scientific realism nor critical rationalism is the argument recognized, that to ascribe 'independent reality' to an object of knowledge is not (to use Kant's expressions) to enlarge the concept of that object, but "to express the relation of the concept to the faculty of knowledge".

The implications of this thesis for scientific realism can be made clear by a brief examination of Harré's account of the logic of existential hypotheses in scientific research. As we have seen, one of the central differences between the methodological theories of realism and deductivism is the role accorded by the former to the postulation of generative mechanisms responsible for producing the phenomena observed by descriptive research, and the subsequent testing of existential hypotheses concerning the entities referred to in iconic models of those generative mechanisms. The 'reality' of such hypothesized entities is confirmed by the verification of existential statements about them, by a form of experimental research which, in an earlier book, Harré referred to as 'ontological experiment' (1961, pp. 42ff.). This notion agrees intuitively with certain basic conceptions of scientific discovery. Thus we know that for a considerable period scientists believed in the existence of phlogiston, holding it to account for certain experimental chemical phenomena. But further experimentation showed that phlogiston did not exist, and that the phenomena of combustion required the postulation of another substance, oxygen, whose existence was then demonstrated. Working with a model of the solar system operating according to the laws of Newtonian mechanics, Adams and Leverrier postulated the existence of a hitherto unobserved planet, which was then shown to exist by the relatively simple experiment of pointing telescopes in the calculated direction. In 1930 Wolfgang Pauli predicted the existence of a hitherto unknown particle, the neutrino, which three decades later, by experimentation with particle accelerators, was also shown to exist.

However, realism as a theory of science requires more than this intuitive agreement with certain episodes of scientific practice. It requires in addition a philosophical account of the nature of these existential claims and ontological experiments. Positivism, according to Harré, is unable to provide such an account, since it does not even accord any role to existence claims and ontological experiments in scientific research. Harré's own account involves a discussion of the criteria used in assessing an existence claim. Such criteria are of two kinds: demonstrative and recognitive. The basic form of demonstrative criterion is spatio-temporal locatability, established by the demonstrative act of ostension. However, Harré's conception of a demonstrative criterion is so wide that the range of possible criteria is extended to include any form of the direction of the attention, which need not even involve a publicly intersubjective act (1970, pp. 67ff.). Thus it is claimed that the existence of a state of mind can be demonstrated by the private report of it by the person whose state of mind it is. This possibility is held to be central to a science of psychology. In the context of the physical sciences, however, 'public ostension to material things and the states and processes in them' is allocated a central role, though this ostension may be indirect, as when the existence of a cause is demonstrated by ostension of its effects (1970, pp. 72ff.). Demonstrative criteria alone are only capable of showing the existence of some, unspecific, thing. In addition recognitive criteria are required, namely criteria referring to the characteristics of the kind of thing to which ostension has been made. Recognitive criteria depend on the kind of existent involved: in the case of things and substances, these criteria must include permanence or duration, since this demarcates things from other categories of being of which existence may be claimed. The recognitive criteria for the properties, powers, and sensible qualities of newly hypothesized kinds of things will be derived from the source of the model in which such kinds of things are postulated: the role of analogy in model construction is to provide a source for describing the characteristics of novel kinds of entities, which then act as criteria by which they may be recognized and their existence verified or falsified (1970, pp. 75ff.).

Brief as this outline of Harré's theory of existential hypotheses is, it is immediately apparent that, from the standpoint of a critical philosophy, the existence of domains of objects is treated by realism as an objective feature of independent reality, rather than as the constituted product of the synthetic activity of the knowing subject. Thus the demarcation of a special domain of material objects, demonstrated ostensively in space and time, and characterized by some degree of permanence or duration, is accorded the status of a metaphysical ontology rather than, as in Kantian critical philosophy, finding its root in the categories through which the manifold of representations is synthesized. Once this transcendental philosophy is given its pragmatist twist, as it was by Peirce and, following him, Habermas, the constitution of the domain of material objects is seen to result from the pragmatic context in which synthetic activity is embedded, namely, the behavioural context of instrumental action. The categories which constitute this domain are not, as we have seen, those of any consciousness whatever, but derive from the historically specific emergence of the human species, and derive their transcendental force from the need of that species to come to terms with its natural environment.

The basic category of the ontology of scientific realism, which is opposed to deductivism's ontology of atomistic events, namely an ontology of powers as the essences of natural kinds, takes on a new significance from the critical standpoint. For, just as there is a hint of recognition of the pragmatic context of instrumental action in Popper's formulation of the independence of reality in terms of 'resistance', so the category of intrinsic powers also points towards the context of instrumental action. To ascribe a power to a thing or material is, as we have seen, to say that in appropriate conditions it will act in a certain way, in virtue of its intrinsic nature (which at a later stage of scientific investigation can be the object of inquiry). If the context of instrumental action is supplied as the framework within which powers ascriptions gain their meaning, then the intrinsic powers of kinds of objects can be interpreted as the intrinsic limits within which human manipulation and technical transformation of such objects is possible. A thing cannot be made to do what it is intrinsically incapable of doing, i.e. what it does not have the power to do. Hence, despite the difference in logical form between scientific realism's construal of natural laws as descriptions of the real essences of things in virtue of which they have the powers they do, and critical rationalism's construal of natural laws as universal propositions based on conjunctions of events, there remains a fundamental similarity once the pragmatic context of the natural sciences is supplied; they point toward the limits of possible technical action in terms of the intrinsic resistance offered by nature to human action. We may remember that Popper himself formulated the notion of the 'technological form' of a natural law, the idea that

every natural law can be expressed by asserting that *such and such a thing cannot happen* (1961, p. 61).

Similarly, the scientific realist's analysis of a powers ascription may also be given a 'technological form', namely as a statement that such and such a material cannot be made to act in a certain way, in virtue of that material's intrinsic nature. But rather than this being given a metaphysically realist interpretation, from the standpoint of critical theory it reveals that, in the methodological categories of the natural sciences, nature has been constituted from the standpoint of an inherent human interest in the possibilities, and also the impossibilities, of its technical manipulation.

The foregoing critique of metaphysical realism, as it is propounded by both critical rationalism and scientific realism, results in the following assessment of scientific realism from the point of view of critical theory: scientific realism, with its conception of the aims of the theory of science in terms of the provision of a rational reconstruction of actual scientific practice, legitimated only by an inadequate doctrine of metaphysical realism, is in fact a new form of 'objectivism'. It fails to raise the problem of the constitution of realms of facts out of the undifferentiated manifold by means of synthetic activity, and hence fails to reflect on the underlying cognitive interest which gives meaning to the methodological framework of the natural sciences. In advancing a descriptive model of natural scientific methodology, which is supposed to do justice to the practical activity of working scientists, it is incapable of comprehending the natural sciences within the wider framework of a theory of knowledge, which would be able to explain why that methodology might (or might not) be an appropriate one for the study of the natural world for a species which has the empirical characteristics possessed by the human species (Habermas, 1972, pp. 195ff.; cf. Hooker, 1975). Being thus trapped within the 'objectivist illusion' identified by Habermas, scientific realism must be judged as subject to critical theory's critique of positivism; in fact, it might be argued, as was hinted earlier, that scientific realism is in this respect even more positivistic than critical rationalism, since it does not reach even the level of reflection of the latter's methodological conventionalism. The partial transcendental reflection which Wellmer and Schnädelbach identified in Popper is not even begun in scientific realism, which believes itself to have described the only possible scientific methodology which is in accord with the real nature of the world.

With this first conclusion in mind, we may now return to the charge of positivism directed at critical theory, specifically at Habermas' theory of the technical interest, by scientific realists. Habermas' epistemological account of the natural sciences, it is argued, remains bound to the deductivist assumptions of positivism, and, as we saw in Section 3.3.2, the very derivation of the technical interest by means of reflection of the methodological categories of the natural sciences depended on the acceptance of a deductivist account of scientific theory and scientific explanation. It is necessary to assume, for the purposes of the present argument, that scientific realism has indeed identified features of modern natural scientific methodology that cannot easily be incorporated within a theory of science based on these deductivist assumptions; it has also been suggested that within the community of philosophers of science this movement away from pure deductivism and towards a realist epistemology, though not necessarily in the form that has been stressed here, is becoming widespread. It must be conceded, I wish to argue, that, despite being bound in its formulation to the illusion of objectivism, realism's critique of deductivism does present difficulties for Habermas' theory of the technical interest. An elaboration of the implications of this thesis for the theory of the natural sciences will conclude this discussion of the interrelationships between the three forms of antipositivism.

It was shown in Section 3.3.2. that two different possibilities were open to thinkers who were convinced of the need to re-open the Kantian tradition of critical philosophy, but were also convinced of the force of the Hegelian and Marxian critique of the ahistorical nature of Kant's theory of the synthetic activity of the knowing subject and of his separation of theoretical and practical reason. The first, derived from Marx and Horkheimer, and pursued in particular by Marcuse and Sohn-Rethel, stressed the historical development of the natural sciences and their relationship to the development of the division of labour, with the implication that the historically universal features of the categories of the natural sciences, if there were any, were less significant than the historically variable and transformable features. The second, advanced in particular by Habermas, concentrated rather on the essential differences between the natural and the social sciences, and stressed rather the historically universal and unalterable pragmatic context of the natural sciences, namely the behavioural system of instrumental action; the resulting theory of the knowledge-constitutive interest in technical control was therefore used to deny the significance of historical variation in the categories of the natural sciences, and to deny the possibility of fundamental transformation of those categories, as was envisaged by Marcuse and Sohn-Rethel.

Although the distinction between these two options seems to be treated by Habermas, in his polemic against Marcuse, as an absolute one, it is possible to reconcile them by interpreting them at different levels of generality. Thus, although the behavioural system of instrumental action may be taken as the universally necessary pragmatic context supplying the framework for any inquiry which would count as a science of nature, this does not exclude the possibility of historical variability, at a still highly abstract but nonetheless more specific level, in the categories defining the way in which a natural science bound to that pragmatic context is realized. This would admit the possibility of quite fundamental changes in scientific methodology, even allowing that in each case knowledge in the natural sciences would remain directed by the knowledge-constitutive interest in possible technical control, and would prompt a reconsideration of the more historically oriented branch of critical thinking on the natural sciences. This is supported by the following straightforward remark of David Bohm:

> a new situation or problem may even require basically new general methods of investigation and general criteria for acceptable theories. For example, consider what is usually called 'the scientific method'. Is it possible,once and for all, to define exhaustively what this is? Evidently not, because this method has itself evolved, and is still evolving, in response to our being confronted with ever new kinds of problems. The scientific method of today contains aspects that were not present several centuries ago, and it seems very likely that in a few more centuries it will be very different again in many respects from what it is now. But this means it must be *continually* changing. Perhaps the change is imperceptible in the short run but, nevertheless, it is clearly a real one (1964, p. 221).

To formulate the matter in that way has the merit of simplicity; however, a thesis of the historical variability of scientific method must be seen, not as the conclusion of one line of thought, but as the beginning of another.

The problem remains that of the analysis of the dimensions of variation of scientific method and of the influences which condition that variation historically. It is only possible to put forward in the present context one or two tentative thoughts along the lines of which research might proceed.

It has been shown in Section 3.3.1. that Marcuse and Sohn-Rethel, following up some hints in Marx, rooted changes in the basic categories of scientific methodology in changes in the division of labour, since productive labour, being the site of human intercourse with the rest of nature, was assumed to be also the site of the synthetic activity in which the realm of nature was constituted and its phenomena ordered under categories. Marcuse was mainly concerned to look towards the future, and drew a crude distinction between a 'repressive' and a 'liberating' mastery of nature, the former corresponding to the form of the division of labour in all present-day advanced industrial societies, and the latter presumably intended to correspond with the form of the division of labour in future, socialist societies. Sohn-Rethel was also concerned to point to this future possibility, but his argument was also designed to explain the origin of the 'mechanistic thought' of modern science stemming from the scientific revolution of the sixteenth and seventeenth centuries, which he found in the expansion of commodity production and the increased division of mental and manual labour. With the help of ideas drawn from scientific realism, it may be possible to state these theses a little more precisely. As we have seen, Bhaskar attributes the basic characteristics of positivist philosophy of science, and in particular its analysis of laws of nature as being or depending on conjunctions of events, to the assumption in that philosophy of a 'closure', the assumption, that is, that the world is a closed system in which constant conjunctions of events occur, and which is therefore regularity-deterministic. According to realism, this is a false assumption, and 'closures' are in fact artificially produced in the laboratory by scientists in order to identify the mechanisms and powers postulated in their theories. This production of a closure is the isolation of an individual process from irrelevant and disturbing factors. I would suggest that the 'real abstraction', which Sohn-Rethel identifies in commodity production, in which all aspects of the productive process are abstracted out except those defined in terms of exchange-value, can be conceived as a form of 'closure'. In a division of labour based on manufacture of commodities, with exchange between separate private producers, the productive process is artificially 'isolated' from its physical environment by the abstraction of exchange: the only conditions and consequences of the productive process that are relevant to the owner of capital are those measurable in terms of exchange value. The productive process is thereby artificially treated as if it were a closed system. If this is the case, and if the productive process is the location of synthetic activity which is thus shaped by the form of the productive process, then the assumption of a closure in positivistic philosophy of science, as identified by the scientific realist, itself derives from and is consistent with the closure of the exchange abstraction.

However, scientific method cannot be understood merely in relation to the nature of the division of labour of the society within which scientific inquiry is carried on. Even after Schnädelbach's distinction between technical and experimental action has been made, scientific research, at least to the extent that it admits of experimentation, remains a form of instrumental action: "experimental action could be described as reflected technical action" (Schnädelbach, 1972, p. 106), and this reflection takes place through the medium of technical action itself, namely through experimentation. And just as productive technical action, if it is to advance beyond a very primitive level, requires the development of tools and then increasingly complicated machinery, so experimental action requires scientific instruments which have become increasingly complicated as scientific knowledge has progressed. The nature of scientific method is, then, also dependent on the level of development of the 'material forces of inquiry', and this will also affect the appropriateness of philosophical accounts of scientific methodology. Bhaskar argues:

> there is in science a characteristic kind of dialectic in which a regularity is identified, a plausible explanation for it is invented and the reality of the entities and processes postulated in the explanation is then checked (1978a, p. 14).

But the ability of scientists to check the reality of such hypothetical entities and processes depends on the level of development of the material forces of inquiry, and it is only in relatively recent times that these have undergone rapid development and transformation. Extensions of the human senses, both macroscopically (telescopes, radio telescopes) and microscopically (electron microscopes, cloud chambers and bubble chambers, and so on), have perhaps only in this century become so interrelated with the development of increasingly fundamental theory that they can be used to check the existence of the entities and processes postulated in such theories. Only under the conditions of extensive and intensive development of the material forces of inquiry does scientific realism become more than speculation.[9] At an earlier stage in the development of these forces, the basic ontology of positivism, that of atomistic events, and the analysis of laws as based on regularities in sequences of observable events, is far more plausible.

A positivist account of the methodology of the natural sciences can therefore be interpreted as reflecting, on the one hand, a division of labour based on the exchange of commodities, and incorporating a division between mental and manual labour, as Sohn-Rethel argued, and, on the other hand, a relatively low level of development of the material forces of scientific inquiry. In addition to the analysis of a systematic relationship between positivist categories and this form of the division of labour, this proposal is at least also consistent with the kind of historical interpretation of the emergence of such basic categories of science as 'the laws of nature' advanced by Edgar Zilsel (1942a, 1942b) and by Joseph Needham (1969); they see the category as stemming from the development of greater precision in the quantitative rules of thumb used by artisans, under the competitive pressure of petty commodity production in early capitalism.[10] There is thus both systematic and historical support for the thesis of those writers within the tradition of critical theory, such as Marcuse and Sohn-Rethel, that there is a close relationship between a positivistic form of the natural sciences and the structure of domination of commodity-producing societies. Scientific knowledge, in the specific form of knowledge of laws of nature which are conceived of as being or depending on empirical regularities, is transformable into technically applicable knowledge in relatively closed systems; this form of scientific knowledge is generated by and then helps to sustain a form of the division of labour which is simultaneously a structure of domination. This form of the division of labour, or, in more abstract terms, of the relationship of the human species to nature, involves the artificial creation of closed systems within an open world, and the motor of this closure is the private utilization of capital, which abstracts out those aspects of the relationship of humanity to nature which are not expressible in terms of exchange-value.

As we saw earlier, the difficulty with this thesis when advanced by Marcuse and by Sohn-Rethel was the apparent lack of any alternative account of the natural sciences. The notion of an alternative natural science therefore appeared as pure speculation, and Habermas' critique of this speculation appeared to have great force. Any conception of a future form of the natural sciences still retains a speculative character; but with the emergence of scientific realism, even in an 'objectivist' guise, there does exist evidence that scientific method does change, and that Habermas' attribution of universality to 'positivist' features of scientific methodology is not justified. Although any form of natural science is, in general terms, bound to the pragmatic context of instrumental action, this does not exclude the possibility of different forms which knowledge of nature might take.

However, just as it would be wrong to attribute universality to these 'positivistic' features of the natural sciences, so it would be equally wrong to argue, as realists appear to, that a realist account of science captures an essence of science which is in turn universal in character. As I have just argued, realism is essentially a response by philosophers of science to a rapid development of the material forces of inquiry, and the form of scientific knowledge that has been thereby made possible. Realism, however, although it makes it possible to identify the assumption of a closure in positivist philosophy of science and to draw a relationship between positivist science and a specific form of the division of labour, does not itself either reflect or point towards any change in the division of labour, or the relationship of mankind to nature. The scientific practice which scientific realism formalizes and reconstructs is still based on the artificial production of closure in experimental conditions, even if the metaphysical assumption of a universal closure has to be abandoned. Scientific realism is the philosophical reflection of a form of scientific knowledge which continues to isolate entities and processes from their open-systemic contexts, just as the productive process of commodity-producing societies is 'isolated' by the abstraction of exchange. Those writers who look to a form of the natural sciences which will reflect a fundamentally transformed relationship of humanity to nature, and who see this transformation as requiring an end to this isolating abstraction, will not find any more support from realism than from positivism or deductivism. They will continue to look for a more holistic, and a more historical, form of natural science.[11]

5.1. Conclusion to Part II

All three forms of antipositivism in the theory of the natural sciences which have been examined in this part have elements of validity. The critical rationalist attack on the positivism of the Vienna Circle was in itself valid, and produced, especially in Popper's early work, a degree of methodological self-reflection unattainable by logical positivism. But its methodological conventionalism remained bound to the basic assumptions of positivism, and the process of reflection could not be carried through. Critical rationalism is thus equally subject to the antipositivism of critical theory as other forms of objectivism. Objectivist theories of science fail to attend to the problem of the constitution of domains of objects of knowledge by the synthetic activity of the knowing subject, and critical theory re-opens the Kantian tradition of critical philosophy, but incorporating the materialist and pragmatist twists of Marx and Peirce. In doing so, however, there is a danger of uncritically accepting the descriptive methodology of positivism while criticizing its objectivism; scientific realism, while itself an objectivist theory, points to the historical variability of scientific methodology by means of a critique of positivist accounts of scientific methodology, a critique which critical theory can accept, extend, and turn to its own purposes.

PART III

ANTIPOSITIVISM IN THE PHILOSOPHY OF THE SOCIAL SCIENCES

CHAPTER 6

THE ANTIPOSITIVISM OF CRITICAL RATIONALISM

In Chapter 2 it was shown that Karl Popper's notion of the nature of the philosophy of the natural sciences was from the outset directed critically against the logical positivism of the Vienna Circle; it was, as far as it went, an essentially antipositivistic conception in that sense. Popper's antipositivism was not always fully understood, and many writers, including some logical positivists, continued to think that there was more held in common between him and the Vienna Circle than there was that divided them. In order to rebut this view, Popper could repeatedly remind readers that his proposal to demarcate science from metaphysics on the basis of the former's falsifiability-in-principle was not, and was never intended to be, a theory of meaning, as logical positivism's principle of verifiability was, and was not designed to demonstrate the meaninglessness of metaphysics. In this way, Popper's antipositivism could be clearly and precisely delineated. However, in the course of this protestation of the purity of his antipositivism it was quite unnecessary to refer to his views on the social sciences and their methods. It is only since the *Positivismusstreit*, and the consequent emergence of the term 'positivism' as a shibboleth, that critical rationalism has had to define specifically its antipositivism in the context of the philosophy of the social sciences.

Throughout the development of the *Positivismusstreit*, the critical rationalists consistently continued to identify 'positivism' in terms of the doctrines of logical positivism, and to reject any extension of the term which would bring critical rationalism within its field of application. Curiously, they also tended to write as if they believed that their opponents, the critical theorists, and in particular Adorno and Habermas, were also using the term in the same way; it was thus possible to argue that the application of the term to critical rationalism betrayed either the critical theorists' ignorance of its opposition to the central doctrines of logical positivism, or else their misunderstanding of the issues at stake. It is worth taking a moment to analyse some of these essentially polemical remarks, which would not otherwise justify serious attention, because they contain an interesting equivocation which lies at the heart of critical rationalism's antipositivism.

In Popper's opening lecture on 'the logic of the social sciences', at the meeting in 1961 in Tübingen which proved to be the beginning of the *Positivismusstreit*, the seventh thesis contains a restatement of Popper's 'critical method', the method of trial and error, of proposing and testing tentative solutions to problems. The critical method is then contrasted with another method which, it is claimed, has been widely accepted in the social sciences;

> the misguided and erroneous methodological approach of naturalism or scientism which urges that it is high time that the social sciences learn from the natural sciences what scientific method is. This misguided naturalism establishes such demands as: begin with observations and measurements; this means, for instance, begin by collecting statistical data; proceed, next, by induction to generalizations and to the formation of theories. It is suggested that in this way you will approach the ideal of scientific objectivity, so far as this is at all possible in the social sciences (Adorno *et al.*, 1976, pp. 90f.).

This view is then identified as the 'myth of the inductive character of the methods of the natural sciences', and is, of course, rejected on familiar grounds. It is also identified as 'positivism'; in a footnote added to the English edition of *The Positivist Dispute in German Sociology*, Popper suggests that

> (w)hat my Frankfurt opponents call positivism seems to me the same as what I here call 'misguided naturalism'. They tend to ignore my rejection of it. (Adorno *et al.*, 1976, p. 91*; cf. Adorno *et al.*, 1969, p. 107).

Leaving until later a discussion of what critical theorists do in fact mean by positivism in this context, what is curious about Popper's disavowal is that it remains unclear, from this passage alone, whether Popper means to reject as 'misguided' any kind of 'naturalism', i.e. any view that the methods of the social sciences are essentially the same as those of the natural sciences, or whether, on the contrary, only this particular, inductivist, and to that extent positivist, form of naturalism is thought to be misguided. The same equivocation appears in Popper's later comments on the *Positivismusstreit* book which have already been quoted:

> the fact is that throughout my life I have combatted positivist epistemology, under the name 'positivism' . . . I have fought against the aping of the natural sciences by the social sciences, and I have fought for the doctrine that positivistic epistemology

is inadequate even in its analysis of the natural sciences which, in fact, are not 'careful generalizations from observation', as it is usually believed, but are essentially speculative and daring (Adorno *et al*., 1976, p. 299).

Here again, it would not be completely clear from this passage alone whether Popper is indeed opposed to the 'aping' of the natural sciences by the social sciences as such, or only to the aping of a specifically positivist version of the methods of the natural sciences.

This equivocation only appears when Popper feels called upon to disavow the appellation of positivist. From his other, earlier and later writings, it is completely clear which of these theses he actually holds. Thus in *The Poverty of Historicism*, Popper proposed:

a doctrine of the unity of method; that is to say, the view that all theoretical or generalizing sciences make use of the same method, whether they are natural sciences or social sciences . . . The methods always consist in offering deductive causal explanations, and in testing them (by way of predictions). This has sometimes been called the hypothetico-deductive method (1961, pp. 130f.).

This doctrine of the unity of method does not, of course, exclude the possibility of more specific differences in method between natural and social sciences, but there are such differences even between different natural sciences. Nor does it exclude the possibility of 'historical sciences', in which the focus of interest is on particular facts describing specific events and not on universal laws and generalizations, but again this does not constitute a difference in principle between natural and social sciences. Since, however, the description of the method, hypothetico-deductive or falsificationist, corresponds to that proposed in *The Logic of Scientific Discovery*, where it was advanced in opposition to the inductivist methodology supported by logical positivism, and since Popper and other critical rationalists identify positivism with, specifically, logical positivism, Popper can claim that this doctrine of the unity of scientific method is an antipositivist doctrine, and that those who claim otherwise either are ignorant or have failed to understand.

The polemical, one might say dissembling, force of this equivocation is quite clear. In the writings of the *Positivismusstreit*, Popper's Frankfurt opponents show themselves to be perfectly well aware of his own rejection of the *misguided* naturalism or scientism of logical positivism, and, as far as it goes, they share that rejection. Thus, although the points of agreement between Popper and Adorno noted by Dahrendorf (1976, pp. 124ff.) did have the curious effect of disguising the extent of their mutual opposition, Adorno's "agreement with Popper's critique of scientism" (Adorno *et al.*, 1976, p. 108, cf. pp. 22ff.) is far from being merely ironic. Popper's critique of the logical positivism of the Vienna Circle is both well known and appreciated by all of the critical theorists who have contributed to the debate, and is clearly enough expressed in the writings to which Popper takes such polemical exception. Hence it would appear as if Popper, and other critical rationalists, could not but be aware that the term 'positivism' had taken on a wider field of applicability in discussion subsequent to Popper's critique, in the 1930s, of the Vienna Circle, that the significant differences between logical positivism and critical rationalism were understood by critical theorists, and that what was under debate was not the doctrines of the Vienna Circle in particular, but, among other things, the tenability of any doctrine of the unity of method of the natural and the social sciences, Popper's included.

In addition, Popper's emphasis on the antipositivism of his views on the methodology of the social sciences was marked by a degree of polemical selectivity. It is of course the case that throughout his life Popper has "combatted positivist epistemology, under the name 'positivism'", but in the context of philosophical discussion of the social sciences this is only a small part of the story. Popper's best known writings on the social sciences, *The Poverty of Historicism* and *The Open Society and its Enemies*, do indeed incorporate antipositivist arguments, but the main thrust of their critical enterprise is directed not against positivism but against 'essentialism', 'historicism', and 'holism'. Against these mistaken doctrines, which are held to characterize the social thought of, for example, Plato, Hegel, Marx, and Mannheim, Popper advances his doctrine of the unity of scientific method and its associated form of scientific politics, 'piecemeal social engineering'. Against that background it is hardly surprising that Popper's critics should find that the similarities between his doctrines and those of unequivocal positivists, be they Comtean or logical, were more significant than the differences, and feel entitled to extend the range of application of the term to him. Nor were the critical theorists of the Frankfurt School the first to do so (Milligan, 1948/49).

The effect of this polemical equivocation and selectivity has been to conceal the real issues involved, and it is to them that attention must now be turned. For this it is necessary to return to the level of the argument of *The Logic of Scientific Discovery*. In that book, it will be remembered, the critique of logical positivism ran along two lines. Firstly, Popper argued against the substantive account of scientific methodology advanced by logical positivism, rejecting induction and replacing it with falsification. Secondly, and more importantly, he criticized the conception of the philosophy of science and of philosophy in general held by logical positivists. He rejected the implications of the application of the principle of verifiability back on to the philosophy of science itself, which would make of philosophy either a branch of logic, or a branch of empirical science, or a kind of anti-philosophizing activity. Instead he wanted to revive a true and independent philosophy in the shape of a theory of knowledge, whose cornerstone was that the distinction between science and non-science could not be found in the nature of things, but rested rather on conventions. Science was to be demarcated from other forms of human knowledge on the basis of its methodological rules, and these methodological rules were to be treated as conventions, advanced and agreed upon because they were most likely to facilitate the purposes set for the gaining of scientific knowledge. There could be no final justification of such methodological rules in terms of their derivation, by deduction, from certain premises, but only provisional justification in terms of their hypothesized suitability for the purpose set; and in the choice of purpose, and hence of method, there remained an irreducible element of human decision. The stress on convention and decision replaced the positivistic search for certainty, and constituted the most significant component of Popper's antipositivism at that time.[12]

We are now in a position to raise the question of the relationship of this theory of knowledge, which has been termed above 'methodological conventionalism', to the doctrine of the unity of method, which Popper first hinted at in his paper 'What is dialectic?' and then expounded in *The Poverty of Historicism* (1969, pp. 321ff.; 1961, pp. 130ff.). In order to be consistent with the methodological conventionalism of the earlier work, Popper would have had to propose a doctrine of the unity of method of the natural and the social sciences as a convention. The convention, that the methodological rules of the social sciences should be the same as those adopted in the natural sciences, would have been supported by an argument that those methodological rules were the ones best suited to pursue the purposes set for the social sciences. These purposes would have to be the subject of decision, and would either be the same as those set for the natural sciences or, if different, pursuable nonetheless by the adoption of the same methodological rules.

The argument in *The Poverty of Historicism* did not, however, take this course. Instead the argument there is, in outline, as follows. In the past, both the advocates and the opponents of a doctrine of the unity of method, pro-naturalists and anti-naturalists, have mistakenly produced doctrines which in various ways exhibit elements of 'historicism', because they have had a mistaken conception of the method of the natural sciences. They have, that is, based their pro-naturalistic or anti-naturalistic arguments on inductivistic and positivistic views of the natural sciences, and therefore either accept or reject the unity of method for the wrong reasons. When the correct view of the methods of the natural sciences is advanced, namely Popper's own falsificationist methodology, it may immediately be seen that this is also a correct description of the methods of the social sciences, as they are described, for example, by Hayek (Popper, 1961, pp. 136f.). It would appear, from this brief summary, as if the methodological conventionalism of Popper's earlier work has now disappeared, and that there has been a relapse into precisely that which was earlier criticized as positivistic; for the doctrine of the unity of method, rather than being proposed as a convention, is advanced as a correct description, either of the methods in fact in use, or perhaps of the 'essence' of both the natural and the social sciences.

This relapse into positivism, as Popper himself had earlier understood it, is most clearly evident in his discussion of sociological laws. In *The Logic of Scientific Discovery*, the idea that there exist in nature invariant regularities, the idea, that is, of the uniformity of nature, was designated metaphysical, since it is an idea that is not falsifiable by empirical evidence. The metaphysical doctrine was replaced by the proposal of a convention:

> the principle of the uniformity of nature is here replaced by the postulate of *the invariance of natural laws*, with respect to both space and time. I think, therefore, that it would be a mistake to assert that natural regularities do not change. (This would be a kind of statement that can neither be argued against nor argued for.) What we should say is, rather, that it is part of our *definition* of natural laws if we postulate that they

are to be invariant with respect to space and time; and also if we postulate that they are to have no exceptions (1968, p. 253).

As a methodological rule, proposed as a convention, this *is* the kind of statement for or against which arguments could be advanced; as we have seen above (Sections 3.1.1. and 3.2.), it is the inadequacy of Popper's arguments in favour of this proposal that leads critical theorists to identify his methodological conventionalism as a form of positivism. However, when Popper turned his attention to the social sciences, the invariance of laws ceased to be seen as a methodological rule which had the status of a convention, but became instead an essential component of scientific method as such:

it is an important postulate of scientific method that we should search for laws with an unlimited realm of validity. If we were to admit laws that are themselves subject to change, change could never be explained by laws. It would be the admission that change is simply miraculous (1961, p. 103).

Elsewhere, even the methodological nature of the idea of laws is lost: thus, in *The Open Society and its Enemies*, having made the distinction between 'natural laws' and 'normative laws', or norms, Popper went on to say:

But of course this does not mean that all 'social laws', i.e. all regularities of our social life, are normative and man imposed. On the contrary, there are important natural laws of social life also. For these, the term *sociological laws* seems appropriate (1962, i, p. 67).

But the bald statement that *there are* sociological laws seems to be precisely the kind of statement that was rejected as metaphysical in the earlier work, for it would appear that there is no way in which this could be known.

It might be thought that this apparent relapse into positivism was merely the result of a looseness of formulation on Popper's part. However, the more systematic argument of Hans Albert exhibits the same features. Albert stresses the element of decision in the choice of methodology, and, with explicit reference to *The Logic of Scientific Discovery*, argues that the theory of science cannot be merely a description of the methods actually in use among practising scientists: this would be 'methodological naturalism', as Popper had said, and would be equivalent to the naturalistic fallacy in moral philosophy (Albert, 1968a, pp. 37f.). The decision that has to be made, however, does not arise in the form of a choice between different proposals for conventions over methodological rules, but in the form of a choice between dogmatism and critical reason: science is the search for the invariant structures of reality, and the method of critical testing is the only one capable of generating progress towards theories with greater degrees of explanatory power (Albert, 1968a, p. 47). There is thus only one set of methodological rules capable of promoting progress in empirical sciences, and these must be essentially the same in both the natural and the social sciences. The demands of the critique of methodological naturalism are satisfied, it seems, by representing the commitment to critical reason as a choice, even though no alternative methodological conventions are seriously on the agenda. In recognizing the element of decision in the acceptance of scientific methodology, one is avoiding the naturalistic fallacy.

The justification for the doctrine of the unity of method of the natural and the social sciences advanced by critical rationalism does not, therefore, seem to be satisfactory even by the standards of Popper's original theory of scientific method. This state of affairs may be understood in the light of critical theory's critique of methodological conventionalism presented above. It was argued by Schnädelbach that methodological conventionalism in the theory of the natural sciences is misleading, because it projects the image of a range of choice between different sets of methodological conventions which cannot exist. The demands of the practical applicability of natural-scientific knowledge place severe constraints on the range of conventions upon which empirical natural science can be based. Methodological conventionalism has the effect of representing as open to decision something over which there can in fact be very little choice; it is a 'sham decisionism'. As such it is hardly surprising that methodological conventionalism reverts to what Schnädelbach calls 'a positivism of the prevailing norms of science' (1971, p. 166), in which the currently prevailing procedures of natural science come to play the role of a new 'positive', since no other procedures are available to be chosen.

The situation in the social sciences, according to the critical theorists, is quite different, for the availability of alternative sets of methodological rules is in their case not just an abstract possibility, but rather the actual state of affairs. A fuller consideration of this will be given in the next chapter, but the point may be initially outlined here. As Habermas points out, at least one alternative set of methodological rules is presented to the social scientist by the existence of the range of 'historical-hermeneutic' sciences, the *Geisteswissenschaften*, such as history, philosophy, or jurisprudence (Habermas, 1970a, pp. 71f.; cf. Kreckel, 1975, p. 98). Philosophical discussion of the methods of these disciplines, namely the interpretation of texts and the transmission of culture, has at least as good a pedigree as the analytical philosophy of the natural sciences, and these discussions are 'no less articulated nor carried on at any lower level' than discussions in the philosophy of the natural sciences (Habermas, 1970a, p. 186). The question of the methodological differences between the natural and the cultural sciences, even if it is now, as Habermas claims, no longer topical, was at one time urgently debated, and opened up the issue of which of these methodologies was more appropriate to the social sciences. Prima facie, therefore, the material exists for the kind of philosophical argument and decision-making which is envisaged by the methodological conventionalism of critical rationalism; there are indeed available alternative methodological rules between which social scientists have to choose, the choice depending, as Popper argued in *The Logic of Scientific Discovery*, on the purpose or aims set for the acquisition of scientific knowledge. If the decisionism of methodological conventionalism in the context of the natural sciences is indeed to some extent a sham, there exists the possibility at least in the social sciences that it could be perfectly genuine.

The possibility of giving serious consideration to alternative sets of methodological rules for the social sciences is, however, by-passed by critical rationalism, which instead opts directly for the doctrine of the unity of method. That this is so reinforces the case made by the critical theorists, and outlined above, that methodological conventionalism was an inadequate response to the problems of methodological naturalism even in the case of the natural sciences. For the programme of adopting the most rationally appropriate methodological conventions in order to pursue freely chosen purposes could not, according to critical theorists, be a satisfactory description of the process of giving epistemological grounding to the methodology of the natural sciences. The relationship between the aims set for science and the methodology proposed by Popper was not one of ends and means, but rather, as Wellmer argued, a conceptual relationship: the methodological rules were a translation of the necessity that knowledge in the natural sciences should be practically (technically) applicable into methodological terms. In addition, the possibility of successful technical application was given foundations in a doctrine of metaphysical realism: technology's increasing success was seen as proof that scientific theory had approached closer to the structure of reality. The combination of these two arguments leads to a vitiation of the claims of methodological conventionalism; for it implies that the only choice is between methodological rules which will facilitate the discovery of theories that approach more closely the structure of reality, and methodological rules that will not. Once the problem is posed in those terms, there is no longer any choice to be made, and the implications of this for the social sciences must be a doctrine of the unity of scientific method; no other method could claim to be studying 'reality'. The *Geisteswissenschaften*, however, also claim to study 'reality', albeit a reality differently constituted from that studied by the methods of the natural sciences. There is, then, a 'choice' to be made, one that is disposed of too quickly by critical rationalism.

The conclusion must be that critical rationalism's adoption of the doctrine of the unity of method of the natural and the social sciences is the result of a failure to carry through consistently the critique of logical positivism's lack of a theory of knowledge. From the perspective of critical theory, this failure is only to be expected, since it views that critique of logical positivism as only a partial one, however accurate. Critical rationalism was a recognition, against the radical empiricism of logical positivism, of the need for a theory of a priori elements in scientific knowledge, but methodological conventionalism was the outcome of an only partial process of reflection which such a theory would require.

The antipositivism of critical rationalism in the context of the philosophy of the social sciences thus turns out to be somewhat one-sided. Its claim to antipositivism is justified insofar as it concerns descriptive methodology: critical rationalism is opposed to the adoption of inductivist methods just as much in the social sciences as in the natural sciences. On the other hand, its acceptance of the doctrine of the unity of method is an inadequate application of its methodological conventionalism even on its own terms, and to this extent critical rationalism fails to do full justice to its own antipositivism.

This assessment does not, however, complete the analysis of critical rationalism's opposition to positivism in the social sciences. This opposition to positivism reveals itself also in the approach taken by critical rationalists to two other sets of problems which arise in forms specific to the social sciences: the problem of values, and the problem of meaning. In both cases doctrines are identified as positivist in terms of their (actual or imputed) relationship to the assumptions of logical positivism, rejected on those grounds, and alternatives proposed. It will be useful to conclude this section with an outline of the relevant arguments in these two areas.

6.1. Critical Rationalism and the Problem of Values

Just as on other topics, critical rationalists identify the 'positivist' theory of values to which they are opposed as that advanced by logical positivism. This theory was a consistent component of the logical positivist attack on metaphysics. It was assumed that value-statements could not be analyzed as functions of atomic statements whose truth-values could be verified by sense-observation, and, if this were the case, it followed from the principle of verifiability that value-statements must be meaningless (see Carnap, 1935, p. 26). As such, value-statements were to be rejected, just as all metaphysics was to be rejected. However, the rather extreme nature of this proposal was apparent to logical positivists, and it appeared necessary to give some account of the nature of value-statements, in particular of ethical and aesthetic judgements which were in widespread use not only by misguided philosophers. This account was the emotive theory of ethics; thus Ayer summarizes his 'critique' of ethics in this way:

> We shall set ourselves to show that in so far as statements of value are significant, they are ordinary 'scientific' statements; and that in so far as they are not scientific, they are not in the literal sense significant, but are simply expressions of emotion which can be neither true nor false (1962, pp. 102f.).

The communicative point of making statements of value might be merely to express, emotively, one's attitude on a certain matter, but it might also be an attempt to influence other people's attitudes on that matter. Ayer noted that disputes about matters of value do occur, but suggested that such disputes often concerned questions of fact, such as whether a particular action was really a member of some

> class of actions by which a certain moral attitude on the part of the speaker is habitually aroused (1962, p. 21).

Questions of fact of this kind were as decidable as any other; if, however, the dispute is not of this factual kind, if, for example,

> the other person persists in maintaining his contrary attitude, without however, disputing any of the relevant facts, a point is reached at which the discussion can go no further (1962, p. 22).

Once the possibility of factual disagreement has been removed, the value-attitudes that remain are essentially undecidable: there is no evidence that could possibly count for or against them.

In his exposition of the doctrines of critical rationalism, Hans Albert sees in this positivist theory of values a radicalization of the difference of 'is' and 'ought', a radicalization which derives from the discovery that belief in an autonomous realm of values is an illusion, and which forms the shared background to both positivism and existentialism (1968a, p. 58). The realm of 'is' is the preserve of scientific knowledge, characterized by objectivity and rationality; the realm of 'ought', on the other hand, is the preserve of radical subjectivity, values being the object of arbitrary decision. The only difference between positivism and existentialism is that adherents of the former find rationality philosophically interesting, while adherents of the latter find irrationality philosophically interesting. Their common presupposition, of a fundamental dichotomy of knowledge and decision, rests however on a mistake. The positivist conception of knowledge as grounded in a basis of certainty is rejected by critical rationalism, as has been seen, and replaced by a stress on the conventional aspect of the methodological rules according to which knowledge is pursued. Hence, "in the last analysis, 'behind' all knowledge there lie decisions" (Albert, 1968a, p. 59); this has two consequences for the positivist account of values: in the first place, it removes the basis for the sharp dichotomy of scientific knowledge and irrational values, for it turns out that in both spheres the role of decision is essential, and in the second place it suggests that the positivist (and existentialist) stress on the arbitrariness of value-decisions is misplaced, for critical rationalists believe that the decisions concerning the choice of methodological rules in science are reasoned ones, and therefore that decisions in other realms can be reasoned too.

The implications of the first of these two points can be seen in critical rationalism's view of value-freedom in science. For logical positivism, value-freedom was, as it were, part of the essence of science; if value-statements were metaphysical, the necessity of their exclusion from science followed automatically. This line of argument is closed to critical rationalism by its rejection of such an essentialist (or naturalist) demarcation of science from metaphysics. If critical rationalism is to adopt value-freedom in science, it has rather to be as a convention. Albert distinguishes between three levels at which values may enter scientific discourse: the level of the *objects* which the statements of science describe; the level of the *object-language*, in which scientific statements about those objects are made; and the level of the *meta-language*, in which for example methodological statements about science can be made (1968a, p. 63). The social sciences raise different problems from the natural sciences, according to Albert, only at the first of these levels; for the subject-matter of the social sciences, unlike that of the natural sciences, consists in part of the values of the individuals under study. Social science cannot avoid making values the object of study. This does not mean, however, that value-statements have to enter the object-language of the social sciences, for statements about values can be of a purely descriptive character. To exclude value-statements from the object-language of the social sciences requires the decision to exclude them, namely, the decision to adopt value-freedom in the object-language as a convention. This decision is itself governed by values, but this is no different from the decision to adopt any other methodological rule; thus values are ineliminable from the meta-language in which methodological problems are raised and discussed, but this is equally the case in both the natural and the social scences (1968a, pp. 63ff.).

If value-freedom in the object-language is to be considered in this way as a convention or, as Albert expresses it, as 'a methodological principle' (1968a, p. 62), then its adoption raises the twin questions of desirability and possibility. The desirability of value-freedom is dependent on the goals set for science, in the same way as with other methodological conventions; but, following a line of argument parallel to that employed in other situations, Albert short-circuits a discussion of the possibility of alternative goals for social science by recourse to the doctrine of metaphysical realism. If the goal set for science is knowledge of the structure of reality, then value-freedom in the object-language is an appropriate convention to adopt (1968a, p. 64). The possibility of value-freedom is dependent, however, on the nature of 'reality', for the elimination of valuation from the terminology of science is made possible by the existence of laws with a universal sphere of validity. There can be no point in forming value-judgements about a universal law, for valuation presupposes the existence of alternative possibilities, which is ruled out ex hypothesi by the universality of a law (cf. Kreckel, 1975, pp. 76ff.). In this way, the discovery of universal laws 'neutralizes' the intrusion of values into the object-language. It remains, however, as yet an open question whether any universal laws of society exist to be discovered by social scientists. If this open question is prematurely closed by an insistence that social reality, like natural reality, must be governed by universal laws, then critical rationalism's doctrine of value-freedom becomes just as essentialist as that of the logical positivists, and the conventional aspect of the principle in effect disappears.

The second point of departure from the positivism of the logical positivists which was derived from the discovery of the essential role of decisions even in scientific knowledge was the doubt cast on the arbitrariness of such decisions. Albert therefore argues against the position held by, for example, Max Weber, and which as we saw was also a component of logical positivism's theory of values, that value-decisions must ultimately come down to irrational and unarguable choices. The existence of any sphere from which critical argument is excluded runs counter to the basic principles of critical rationalism, and while Albert recognizes the possibility of such 'dogmatization' of values, he disputes its necessity. Critical rationalism's opposition to positivism on this point is a rejection of 'ultimate presuppositions' which would be immune to all critical discussion, and not open to the influence stemming from the findings of social scientific research. Albert is therefore not prepared to accept that the gap between 'is' and 'ought', between knowledge and decision, is as wide as positivistic presuppositions would have it be. Knowledge can influence decision, value choices are never inherently unarguable (1968a, pp. 68ff.; cf. Runciman, 1963, p. 173). It must be stressed, however, that this opposition to positivism does not extend to a complete abandonment of the distinction between fact and value. It remains the case for critical rationalism, just as for logical positivism, that value-judgements cannot be derived directly from descriptions of states of affairs; the logical gap between fact and value, which makes it impossible for value-statements to be true or false, is upheld by critical rationalism no less strongly than by logical positivism, whatever other criticisms of 'positivist' theories of values are made. In this sense, both doctrines are united on the illegitimacy of a science which goes beyond the criticism of scientific statements to the criticism of the object of those statements, of social reality itself.

6.2. Critical Rationalism and the Problem of Meaning

Given the commitment of critical rationalism to the doctrine of the unity of scientific method, its proponents are in no position to view the meaningfulness of social action and of human culture as the basis for any form of dualism in the sciences. In this they are at one with the logical positivism which they otherwise criticize. Both reject the claims put forward for the human or social sciences, or the *Geisteswissenschaften*, to a method or an aim which diverges markedly from those of the natural sciences. Both deny that the understanding of meaning, or *Verstehen*, can constitute either an alternative source of scientific knowledge which is appropriate to the study of human action and culture, or a goal for such knowledge different from the goal of causal explanation in the natural sciences. Since the *Positivismusstreit*, however, critical rationalists have been under somewhat greater pressure than logical positivists ever were to respond to the challenge of those who have made the meaningfulness of human culture the foundation of a dualism of the natural and social sciences, and both Popper and Albert have written more extensively than hitherto on the problems of meaning and understanding in the sciences. In these more recent writings they have outlined approaches to the problem of meaning and understanding which are to be consistent with the doctrine of the unity of scientific method, and have also advanced criticism of their dualistic opponents. In the present context, their criticism of 'methodological separatism' (Albert, 1968a, p. 131) is more relevant than their positive proposals, since in the course of that criticism the term 'positivism' has been extended by critical rationalists to include certain doctrines which, it is claimed, are held by methodological separatists. The basis for this new charge of 'positivist' will here be outlined; no extensive consideration will be given to the positive side to critical rationalists' theories of meaning, either to Popper's theory of the 'third world' or to Albert's proposal for a "theory of human interpretive activities" which would assimilate *Verstehen* to nomological science by providing causal explanations of the interpretive behaviour of human beings (Popper, 1972, pp. 153–190; Albert, 1968a, pp. 152ff.).

Albert identifies two traditions which have each laid particular stress on the problem of meaning in the social sciences: that of historicism and hermeneutics, and that of Anglo-Saxon analytic philosophy. The former is characterized by its rejection of two modes of procedure on the grounds of their ahistorical nature: namely of the application to historical phenomena of absolute and timeless standards of judgement, and of the explanation of the course of history and hence of social and cultural phenomena by means of universal, timeless laws, on the model of the natural sciences. Instead, according to Albert, historicist hermeneuticists

consider the historical process as a system of events, which bear their meaning in themselves, and which should be grasped in their context of meaning directly, i.e., without the application of universal standards or laws. In the place of natural scientific explanation there should appear a method of interpretative understanding which grasps the unique and the individual, an individualizing method, which should make the essence of the phenomena intelligible in relation to their historical development (1968a, p. 133).

In criticism of this programme, Albert fastens on the 'directness' or 'immediacy' of the experience which is supposed to be at the root of this method, and condemns it as 'positivist': it is

a programme in which, in a similar way to that in the positivism of the natural sciences, immediate experience was stressed as a means of grasping the given, and which in both cases was connected with a strong mistrust of theoretical conceptions. *Hermeneutic positivism* in the theory of the *Geisteswissenschaften* ... is in this respect to be considered as a parallel phenomenon to the *sensationalist positivism* which at one time had a certain influence on the development of the natural sciences ... Both philosophical tendencies accentuated direct, pre-theoretical experience as a means to knowledge; only in the one case it was 'outer', in the other rather 'inner' experience, or, otherwise expressed, whereas the one wanted to fall back above all on the experience of the senses (*Sinneserfahrung*), the other laid greater value on the experience of meaning (*Sinnerfahrung*), which in the comprehension of socio-culturally significant relationships seemed to play an analogous role (1968a, pp. 133f.).

This 'myth of the given' and of direct experience of it is then rejected in the context of hermeneutics in the same way as Popper had rejected it in the context of sense-experience in the natural sciences.

The second tradition identified by Albert is analytic linguistic philosophy, stemming from the later writings of Wittgenstein. Wittgenstein had recognized that the error in his earlier work, as in that of the logical positivists, was to allow only one use of language, the descriptive use found in the natural sciences, to be considered meaningful. He realized that this use was in fact only one among many, and had no claim to priority in meaningfulness. This opened the way to the analysis of numerous different 'language-games', each of which could be understood as a different social 'form of life'. Albert suggests that the kind of investigation prompted by this initial idea, the study of the logical grammar of particular expressions and statements in actually existing languages, could be pursued as a scientific inquiry whose findings were open to critical testing. But, he argues, this has not been the outcome. Instead, analytic linguistic philosophy, however meticulous and detailed its researches have been in comparison with the quasi-theological universal ontology into which hermeneutics has developed, has turned into an aprioristic sociology, whose method of intuiting the conceptual structure of language-games is incompatible with the critical method of scientific inquiry. In this connection, Albert (1968a, pp. 146f.) cites in particular Winch's "idea of a social science" (1958). In criticism, he argues that this aprioristic sociology is just as subject to the "myth of the given" as hermeneutics. He quotes Wittgenstein's well-known remark:

> Philosophy may in no way interfere with the actual use of language; it can in the end only describe it. For it cannot give it any foundation either. It leaves everything as it is (1963, p. 49e),

and concludes that it supports a new form of 'positivistic' attempt to conceive of the understanding of meaning as direct access to a given. It is therefore subject to the same objection as any other form of 'positivism': its illusion of the given conceals the theoretical, or conventional, aspect of all knowledge. In addition, Albert argues, this conceptual analysis is to be condemned on account of its conservative consequences: to "leave everything as it is" is to withdraw it from criticism.

> Theory and criticism, explanation and enlightenment disappear together, and give way to a conservative descriptivism which is suited to the unexplained acceptance and uncritical taking-over of whatever already exists, as it is embodied in traditional forms of life (1968a, p. 148).

Both of these forms of 'positivism' are incompatible with the tradition of critical thought.

Two points may be briefly made concerning these criticisms of dualistic or separatist approaches to the social sciences which base themselves on the meaningfulness of social action and culture. Firstly, and polemically, it may be remarked that critical rationalists, who so strenuously reject the extension of the term 'positivist' to apply to themselves, and claim to use it in a definite and highly restricted sense, are themselves by no means loth to extend its range of application when it suits them, and to use it critically against their opponents. Secondly, as will be seen in more detail below, there is a curious degree of apparent correspondence between Albert's criticisms of hermeneutics and analytic linguistic philosophy and those directed against the same tendencies by critical theorists.

CHAPTER 7

THE ANTIPOSITIVISM OF CRITICAL THEORY

I have already briefly mentioned that, in the context of the social sciences, the antipositivism of critical theory has taken on a twofold character: it is at once epistemological and sociological. This dual character can be identified through the two separate meanings of the term 'critique' which plays such an important role in the thought of the critical theorists.[13] In the first place critique derives its sense from the meaning it had in Kant's 'critical philosophy', the attempt to determine the conditions of possible knowledge by reflection of the knowing subject's own cognitive faculties: this is the dimension of epistemological critique, or critical epistemology. In the second place, critique

> denotes reflection on a system of constraints which are humanly produced: distorting pressures to which individuals, or a group of individuals, or the human race as a whole, succumb in their process of self-formation. Critique in this sense has its root in Hegel. In the *Phenomenology of Mind* Hegel developed a concept of reflection which presents the idea of a liberation from coercive illusions (Connerton, 1976, p. 18).

This notion of critique was taken up by Feuerbach, and then finds its form which most obviously constitutes the source of critical theory in Marx's 'critique of political economy'. In the shape of a critique of the specific ideology produced by capitalist society, Marx also criticized the society itself from the standpoint of the 'determinate negativity' of that society, the force capable of transforming it into a society corresponding more closely to the needs and potentialities of the human species. This we may call 'sociological critique', although it is also the case that an influential model for this form of critical reflection has been found in Freudian psychoanalysis.

In the early days of the Frankfurt School, the second of these two meanings of 'critique' was predominant. In his programmatic essay, 'Traditional and critical theory', Max Horkheimer introduced the notion of critical activity with the explanatory remark that

> The term is used here less in the sense it has in the idealist critique of pure reason than

in the sense it has in the dialectical critique of political economy. It points to an essential aspect of the dialectical theory of society (1972, p. 206, n. 14).

Critical theory was conceived as the continuation of the Marxian tradition. Yet even in these early essays the Kantian conception was not absent. Kant was defended against 'the latest attack on metaphysics', that of logical positivism or logical empiricism, for his activist theory of knowledge, his attempt to provide epistemological foundations for science rather than merely taking science as equivalent to knowledge, as the positivists did, and for his idea, however vague and idealist, of a universal subject, a subjective totality, which can be imagined but which does not yet exist (Horkheimer, 1972, pp. 157ff., 202ff.). However, the ambiguities and obscurities in Kant's writings were traced to the contradictions inherent in bourgeois society, and the neo-Kantian dualism of the natural and cultural sciences did not appear to be an appropriate starting-point for an epistemological account of the social sciences. In any case, the main concerns of the critical theorists were not epistemological, but directed rather towards the development of critical theory itself, with the intention of contributing to social change (Jay, 1973, p. 46; Wellmer, 1971, pp. 14f.).

With the shift from the critique of political economy to the 'critique of instrumental reason' in the 1940s, this balance between the two notions of critique underwent change. For, although the point of the critique of instrumental reason remains 'sociological', in that it is an attempt to analyse the forms of domination and its legitimation in advanced industrial societies from the standpoint of the possible sources of change, the substance of the sociological critique increasingly raised epistemological questions. If the central form of ideology which legitimated relations of domination in advanced industrial societies was no longer the ideology of equal exchange of liberal capitalism, but rather the more abstract ideology of technical or instrumental reason, a critique of the social sciences which formed part of that ideology and contributed to the administration of domination in these societies must itself become a pressing task. Initially perhaps for reasons connected with the task of sociological critique, epistemological critique moved higher on the agenda of critical theory. If the social sciences were not to be bound to the technical interest of social administration, as they would if they followed the lead of the natural sciences, then alternative methodologies for them would have to be given epistemological foundations. The contributions of the critical theorists to the epistemology of the social sciences during and after the *Positivismusstreit* have to be understood in this light. The shift towards epistemology could never, however, be a declaration of independence. Even though the writing of the most important of these later critical theorists, Jürgen Habermas, is an explicit return to Kantian themes, the two elements of critique are never completely separated, but are united through the concept of 'cognitive interests':

> The analysis of the connection of knowledge and interest should support the assertion that a radical critique of knowledge is possible only as theory of society (Habermas, 1972, p. vii.).

Unlike that of critical rationalism, therefore, the antipositivism of critical theory operates explicitly both at the level of epistemology and at the level of theory of society.

However, in order that the antipositivism of critical theory may be outlined at a comparable level to that of critical rationalism and of scientific realism, this link between epistemological critique and sociological critique will in what follows be, initially at least, partially dissolved. To the extent that this is possible, the main themes of the epistemological critique of positivism in the context of the social sciences will be put forward first, leaving the sociological critique of positivism until later.

In the case of the natural sciences, as we have seen, critical theory identified positivism as the set of doctrines developed within the complex problematic erected on the basis of a set of epistemological and ontological presuppositions: namely the desire to ground knowledge in a basis of certainty; the identification of this certainty with the certainty of pure sense-experience; the analysis of sense-experience on the model of pure receptivity; and an ontology of atomistic nominalism. This positivism, including those conventionalist reactions which failed to break sufficiently with those presuppositions, was then subjected to a critique from the standpoint of the Kantian critical philosophy; the descriptive methodology advanced by the most progressive form of 'positivism', namely critical rationalism, was accepted as adequate for the natural sciences, but the theory of science in which that descriptive methodology was incorporated was rejected because of its inability to reflect the conditions which made the application of that methodology possible and appropriate. Positivism, according to critical theory, is incapable of developing a reflective theory of the nature of the subject of knowledge and the relationship of the subject to the object of knowledge, in this case, to nature. In fact, as Habermas puts it, "that we disavow reflection *is* positivism" (1972, p. vii).

The case of the social sciences is more complicated. The inability of positivistic philosophy to provide a reflective epistemology of the natural sciences has no effect on their methodological practice; on the contrary, the failure of the natural sciences to reflect their knowledge-constitutive interests protects them from aberrations (Habermas, 1972, p. 315). In the social sciences, positivism has its consequences both at the level of methodology and at that of epistemology. In the first instance, positivism in the social sciences is identified as the unreflectively dogmatic doctrine of the unity of science, which imposes the methodology of the natural science on to the social sciences too. In an extended sense, positivism is also found in the failure of other methodological approaches in the social sciences to reflect their own epistemological conditions, as a result of which they too fall into the illusion of objectivism. Thus positivism can be seen not only in theories of unified science, but also in separatist theories of the methodology of historical and hermeneutic cultural sciences, and in falsely objectivist accounts of critical sciences with emancipatory potential. These three cases will be taken one by one.

7.1. The Critique of the Doctrine of the Unity of Science

As was briefly mentioned in the previous chapter, critical theory has an initial prejudice in favour of a theory of the diversity of the sciences, or methodological separatism. Habermas justifies this prejudice by pointing to the existence of the historical and hermeneutic sciences, whose philosophical basis and methodology has been the subject of just as much disciplined and critical thought as has the theory of the natural sciences. The tradition of the *Geisteswissenschaften*, the theories of hermeneutic understanding going back to the writings of Schleiermacher and beyond (Palmer, 1969; Bauman, 1978), and the historicist tradition in German historiography (Iggers, 1968), have all produced a philosophical literature concerning the status of the claim to knowledge of the humanistic disciplines which cannot be swept away by a hasty decision in favour of the doctrine of the unity of science, and which prima facie have at least as good a claim to provide the foundations for the methodology of the social sciences as the sciences of nature. Even critical rationalists, it was argued above, should find in the existence of the *Geisteswissenschaften* material for a genuine and reasoned choice between sets of methodological conventions on the part of the social sciences. The task for critical theory is no less intricate. If it is to strengthen its initial prejudice against the unity of science into a convincing argument, it has to show that the methodology of the natural sciences is inappropriate for the study of social reality, to investigate the conditions under which the methodology of the historical-hermeneutic disciplines are applicable and hence whether they would be a more appropriate model for the social sciences, and, if not, to develop an alternative basis for the social sciences which follows the lead neither of the natural sciences nor of the humanistic disciplines.

The inappropriateness of the methodology of the natural sciences for the study of social and cultural reality is ultimately derived from what is held to be a fundamental difference in the modes of experience which give access to natural reality on the one hand and social reality on the other. To describe these two modes of experience, Habermas uses the terms 'sensory experience' and 'communicative experience' (1970a, pp. 188ff.).[14] The difference between these two forms of experience appears at various levels, but the basic assumption is that, whereas we can achieve access to information about natural phenomena by means of the experience of sense-perception, access to information about social and cultural phenomena can only be gained by going beyond sense-perception to the method of the understanding of meaning. The experience of social and cultural reality takes the form of the understanding of the signs, symbols, and language used by the participants in the social contexts which form the object of study. These symbolically structured phenomena cannot be understood by sensory experience alone, but only if they can be made intelligible and communicable to other people.

The initial consequence drawn from this difference in modes of experience is that the relationship between subject and object is fundamentally different in the experience of social reality from that of natural reality. For the process of understanding presupposes a relationship of communication in which symbolic meaning can be transmitted and received.

Habermas also describes the understanding of symbolically structured social reality as 'dialogic', in contrast to the 'monologic' understanding of statements describing the results of sensory perception in the natural sphere (1972, pp. 161ff.), and goes so far as to deny that the relationship between a social scientist and the people whose social life he is studying should even be conceived of as a relationship between subject and *object*. Since the understanding of meaning necessarily requires a communicative relationship between the person who aims to understand and the person or people who are to be understood, this relationship should rather be conceived of as an interaction relationship between subject and co-participant (*Gegenspieler*) than one between subject and object (1970a, p. 189). The role of the subject in investigation based on communicative experience is in fact quite different from that of the subject of sensory experience; while the latter may be described as a 'disengaged observer', the former plays rather the part of a 'reflective partner' (*reflektierter Mitspieler*).

It follows that the form of synthetic activity in which the object-realm of the social and cultural sciences is constituted is quite different from that in the natural sciences. We have seen that in the latter case, the 'pragmatist' re-interpretation given to the Kantian conception of synthetic activity located the constitution of the object-realm of nature in the practical activity of labour, or, more generally, 'the behavioural system of instrumental action'. In the case of those disciplines which gain access to social and cultural reality by means of communicative experience, synthetic activity must be located not at the level of instrumental action but at that of 'communicative action': here

> reality is constituted in a framework that is the form of life of communicating groups and is organized through ordinary language. What is real is that which can be experienced according to the interpretations of a prevailing symbolic system (Habermas, 1972, p. 192).

It can then be seen that there is another difference between the constitution of 'observable' natural reality and that of 'intelligible' social and cultural reality:

> the pattern of communicative action does not play a transcendental role for the hermeneutic sciences in the same way that the framework of instrumental action does for the nomological sciences. For the object-domain of the cultural sciences is not constituted only under the transcendental conditions of the methodology of inquiry: it is confronted as something already constituted (1972, p. 193).

In other words, whereas the object-realm of the natural sciences consists of states of affairs which are only interpreted in the course of being investigated, in terms of the categories which structure the investigation, the object-realm of the hermeneutic sciences consists of states of affairs that are pre-interpreted by their own cultural symbolic meaning-systems (Habermas, 1970a, p. 188).

Cultural meaning-systems are essentially *historical* phenomena. They can be conceived as institutionalized values and norms, or as systems of constitutive rules, which have no universal validity but are restricted to particular historical social circumstances, and are subject to historical change (cf. Kreckel, 1972, p. 40). The implication drawn from this is that, unlike natural phenomena, social and cultural reality, so far as it consists of intelligible systems of meanings, cannot be subject to universal and invariant laws; on the contrary, social and cultural reality has been created by human beings, not necessarily with full awareness of their creative activity, in the course of history, and remains always open to further transformation in the future. The category of universal or invariant law, which plays the central role in the philosophy of the natural sciences, according to both Habermas and his 'positivist' opponents, cannot be an appropriate category for the object-domain of the cultural sciences: that which is hermeneutically understandable cannot also be historically invariant.

The dichotomy between the natural sciences, whose mode of experience is sensory, and whose basic category is that of invariant natural laws, and the hermeneutic or cultural sciences, whose mode of experience is communicative, and whose basic category is the historically specific, intelligible system of meaning, does not immediately settle the question posed by the doctrine of the unity of science. Popper too, for example, is prepared to allow a division between theoretical and historical sciences, not on the basis of the mode of experience appropriate to each or on the basis of presence or absence of 'intelligibility', but rather on the basis of the theoretical sciences' interest in the general or universal, and the historical sciences' interest in the particular or specific (1961, pp. 143ff.). Critical theory's division between natural and hermeneutic sciences does not exclude the assimilation of theoretical social sciences, such as sociology or economics, to the nomological model of the theoretical sciences. It would be possible to exclude this as it were by definition, and this is what Habermas at one point appears to do. He states:

> Meaning, which is intended in action and which is objectivated both in language and in actions, is transferred from social facts to relations between facts: there are no empirical regularities in the realm of social action which, even though not intended, would not yet be intelligible (*verständlich*) (1970a, p. 167).

Since any regularity which is intelligible cannot be also invariant, this statement is tantamount to the exclusion of the possibility of invariant laws in social science by fiat. This, however, is unsatisfactory. Hans Albert has put the objection:

> Why should there not be for example universal laws concerning learning, change, and stability of norms, which could contribute to the explanation of social development? Can Habermas know that a priori? (1968b, p. 344.)

and, as Reinhard Kreckel (1972, p. 43) has argued, this objection is convincing. Nonetheless, the intelligibility of social and cultural reality provides the critical theorist with another argument. We may assume that theoretical social sciences, attempting to follow the model of the natural sciences, have in fact discovered regularities in the realm of social action. Even if these regularities are not known to be invariant laws, they may be used, in the same way as invariant laws of nature, to derive technical recommendations for the achievement of goals of social administration, for example the management of the economy, within the specific sphere within which they are known to hold good.[15] However, critical theory is able to advance a criterion for deciding whether such a regularity should be treated as governed by the operation of a universal law (in which case, although it may be used for purposes of technical application, it constitutes an unchallengeable limit to human action in the same way as natural laws), or whether it should be conceived of as a historically specific correlation of circumstances, which is not governed by an invariant law but which is open to transformation by human action (cf. Habermas, 1972, p. 310). This criterion is the 'intelligibility' of the regularity concerned, since we have seen above that social and cultural reality, to the extent that it is intelligible within a framework of cultural meaning, cannot be assumed to be historically invariant, but will instead partake of the historical specificity of cultural meaning-systems in general. Thus,

> An empirically discovered 'constant', the historical specificity of whose emergence and

development can be reconstructed and therefore 'understood', must not immediately be equated with a 'genuine' invariance (Kreckel, 1975, p. 105).

This argument does not rule out the possibility of real invariant laws in the theoretical social sciences by fiat, but subjects candidates to the severe test, that it should not be possible to provide a hermeneutic interpretation of them, before a claim to their universality should be entertained.

If this line of argument is convincing, it provides the basis for an epistemological critique of the doctrine of the unity of scientific method, and opens the door, as will be seen later, to an alternative, critical social science. For 'positivism' is unable to make the distinction between 'genuine' invariant regularities and historically specific, intelligible relationships.

Positivism ... denies the dualism of the sciences as such. It disputes the case that sociology has any connection with history which intrudes into its methodology: there is no genuine access to history at all. Hermeneutics is pre-scientific, and even the historically oriented sciences obey the indivisible logic of unified science, which abstractly relates systems of propositions to the data of experience. Methodology can make no structural difference between nature and history in the mass of appearances (Habermas, 1970a, p. 100).

Positivism is in fact unable to make the distinction between sensory and communicative experience which provides the initial basis for the dualism of the natural and the hermeneutic sciences. Positivists who attempt to come to terms with the method of *Verstehen* tend to interpret it as a process of attaining knowledge of the mental states, or motives, of acting individuals, and, using the standards of objectivity of sense-experience in the natural sciences, condemn *Verstehen* as incapable of producing objective knowledge. They see it as merely 'empathic identification', generating hypotheses which stand in need of controlled testing (Nagel, 1961, pp. 484f.; Abel, 1948). This "positivist misconception of *Verstehen*" (Halfpenny, 1976, pp. 213ff.) rests on a failure to appreciate the dialogic relationship in which communicative experience is necessarily constituted, and hence on an insistence on forcing *Verstehen* into the mould of monologic sensory experience. On the basis of this initial misconception, the assessment of the specific nature of hermeneutic understanding must of necessity be negative (Wax, 1966/67; Leat, 1972; Outhwaite, 1975).

Since they misconceive the nature of *Verstehen*, it follows that adherents of the doctrine of the unity of scientific method lack, in Reinhard Kreckel's felicitous phrase, "a perceptual organ for the intelligibility of empirical relationships" (1972, p. 42). They therefore have no use for, and no way of drawing, the distinction, so important to critical theorists, between invariant laws and regularities which, on account of their intelligibility, must be provisionally judged to be historically specific and in principle alterable. There is a further, and essential, reason why this distinction is of such concern to the critical theorist and of no interest to the 'positivist', the adherent of the doctrine of the unity of scientific method. In the context of the philosophy of the natural sciences, as we have seen, the critique of positivism developed by critical theory is based, in Habermas' terms, on the discovery of positivism's 'objectivist illusion', its belief that knowledge relates directly to objective facts, without the mediation of a categorial framework constituting those facts from the standpoint of a specific interest in knowledge, which in the case of the natural sciences is an interest in possible technical control. Nature is investigated by science under the category of invariant natural laws because this makes possible, in the behavioural system of instrumental action, technical manipulation, or evasive action in the case of practically unmanipulable consequences, according to reliable, because invariant, technical rules. Positivism, caught in the objectivist illusion, is unable to discern this activity of constitution and the human interest in knowledge which it serves, and believes instead that, in discovering invariant laws, science is attaining knowledge of 'reality' as such. It therefore assumes that all scientific knowledge, no matter which aspect of 'reality', natural or social, it concerns, must take the form of invariant laws or hypothetical candidates for such invariant laws.

Once the objectivist illusion has been disclosed, and the connection between the categorial framework in one realm of scientific knowledge and the human species' interest in knowledge in that realm has been established, the possibility is opened up that the categorial frameworks in different realms of knowledge are not associated with the same knowledge-guiding interest. This, as is well known, is the course taken by Habermas. He identifies two other knowledge-guiding interests which structure the categorial frameworks within which knowledge is pursued. The first of these, the 'practical' interest in "the preservation and expansion of the intersubjectivity of possible action-orienting mutual understanding" (1972, p. 310), will be discussed more fully below. For the present, it is more pertinent to outline the second of these two further cognitive interests, the 'interest in emancipation'. The interest in emancipation, which structures the acquisition of knowledge according to the category of 'self-reflection', has as its aim to release the subject of reflection, human beings, "from dependence on hypostatized powers" (1972, p. 310), in order that they may achieve "autonomy and responsibility" (*Mündigkeit*) and realize "the good life" (1972, p.314). Whereas the technical interest is tied to the behavioural system of instrumental action (and the practical interest to the system of communicative action), the interest in emancipation is not restricted to any particular system of human action, but instead, since it derives its validity from the human interest in reason as such, may be followed in any context where the operation of human reason holds sway. Following Hegel, who developed this concept of reflection in the *Phenomenology of Mind*, and Fichte, who developed the concept of an innate emancipatory interest in reason, Habermas describes

> the experience of the emancipatory power of reflection, which the subject experiences in itself to the extent that it becomes transparent to itself in the history of its genesis. The experience of reflection articulates itself substantially in the concept of a self-formative process. Methodologically it leads to a standpoint from which the identity of reason with the will to reason freely arises. In self-reflection, knowledge for the sake of knowledge comes to coincide with the interest in autonomy and responsibility (*Mündigkeit*). For the pursuit of reflection knows itself as a movement of emancipation. Reason is at the same time subject to the interest in reason. We can say that it obeys an *emancipatory cognitive interest*, which aims at the pursuit of reflection (1972, pp. 197f.).

Thus this emancipatory "human interest in autonomy and responsibility is not mere fancy, for it can be apprehended a priori" (1972, p. 314).

Since it is not restricted in its range of applicability to any particular context of action, but is rather related to the interest in reason as such, the relation of the emancipatory interest to scientific knowledge is not the same as in the case of the technical and the practical interests. It is rather an overarching principle, which sets the standards against which the rationality of a methodological framework for a science can be measured (cf. Kreckel, 1972, p. 36). It is therefore epistemologically prior to the specific cognitive interests of the natural and hermeneutic sciences:

> Indeed, the category of cognitive interest is authenticated only by the innate interest in reason. The technical and practical cognitive interests can be comprehended unambigu-

ously as knowledge-constitutive interests only in connection with the emancipatory cognitive interest of rational reflection (Habermas, 1972, p. 198).

In the case of the natural sciences, this means that a methodology which accords with the technical interest is most appropriate because it is in accord with the interest of the human species in rationally emancipating itself from dependence on and subjection to the forces of nature.[16] Emancipation proceeds from the recognition that there exist invariant laws of nature which set limits for human action which are beyond human control and criticism, but which, if they can be discovered and organized into a coherent system, offer the prospect of the greatest possible control over natural processes open to human manipulative ability. But in the case of the social sciences, which take as their object the historically specific and culturally intelligible norms, values, and institutions of societies, emancipation does not take the same form. As we have seen, such cultural formations, because of their historical specificity, must be assumed to be the product, although not necessarily the intentional product, of human activity, and thus, in principle at least, open to continual human criticism and transformation. Emancipation in this sphere need not take the form of a recognition of necessary limits set by invariant laws, but can take the form of the rational criticism and active transformation of irrational or unsatisfactory cultural and social formations with a view to the creation of more rational ones, in which autonomy and responsibility may be achieved and the good life led. This explains why the distinction between 'genuine' invariant laws and historically specific and in principle transformable regularities in the realm of social life, which as we saw above cannot be drawn by adherents of the doctrine of the unity of scientific method, is so crucial to the critical theorist. Habermas puts it like this:

A critical social science ... is concerned ... to determine when theoretical statements grasp invariant regularities of social action as such and when they express ideologically frozen relations of dependence that can in principle be transformed. To the extent that this [the latter – Author] is the case, the *critique of ideology*, as well, moreover, as *psychoanalysis* [the two models of a critical social science – Author], take into account that information about law-like connections sets off a process of reflection in the consciousness of those whom the laws are about. Thus the level of unreflected consciousness, which is one of the initial conditions of such laws, can be transformed. Of course, to this end a critically mediated knowledge of laws cannot through reflection alone render a law itself inoperative (*ausser Geltung*), but it can render it inapplicable (*ausser Anwendung*) (1972, p. 310).

Positivism, unable to make this distinction in any methodologically relevant way, is incapable of imagining a critical social science of this kind. To the extent that positivism recognizes a process of emancipation, it must conceive it as a form of enlightenment, of emancipation from ignorance concerning the actual structure of 'reality', both natural and social; improvement of knowledge of the laws which govern this 'reality' will help humanity come to terms with the limits of what is possible, and will help in the technical manipulation of specific processes taking place within reality. The structure of such knowledge and its relationship to its object is, for positivism, always the same; positivism's failure to develop a critical epistemology not only leaves it trapped in the objectivist illusion of a metaphysical realism, but also makes it impossible for it to imagine differently constituted object-domains in which different methodologies would serve human interests.

7.2. Latent Positivism in the Hermeneutic Sciences

The critique of the positivism of the doctrine of the unity of scientific method can easily be understood as a straightforward continuation of the critique of positivism in the context of the philosophy of the natural sciences. Positivism's lack of a reflective critical philosophy denies it the possibility of advancing an epistemological theory which would specify the conditions under which the methodological rules of the nomological sciences were and were not appropriate. Critical theory's antipositivism has further ramifications, however, and includes within its scope what it sees as mistakenly positivistic theories of both the hermeneutic and the critical sciences. In the case of the hermeneutic disciplines, it is first necessary to outline what the critical theorist identifies as their epistemological basis and to raise the question of the applicability of their methodological rules to the social sciences, before specifying what antipositivism means in their case.

The object-realm of the hermeneutic sciences, namely the linguistically and symbolically structured systems of cultural meanings, is constituted, as we have seen, in the system of communicative action, as opposed to the object-realm of the natural sciences, which is constituted in the behavioural system of instrumental action. The former object-realm therefore appears as already constituted before the emergence of procedures of systematic investigation with their own categorial schemes. Communicative experience is one of the two basic forms of experience of the human species, and derives from its character as a cultural species, the members of which communicate with each other, and therefore need shared media of communication which make possible mutual understanding and intersubjectivity. The cultural meaning-systems, or cultural traditions, which come to form the subject-matter of the cultural sciences, serve the purpose of making possible meaningful interaction and communication between individuals who are members of one species, but whose experiences differ considerably from one another. In order that interaction may be based on mutual understanding, and that individuals may communicate about their experiences, it is necessary that they should be able to interpret the expressions which others make within the context of systems of cultural meanings. This is the primary location for the method of *Verstehen*, or of hermeneutic interpretation.

Habermas argues that the specific features of hermeneutic interpretation can be seen by showing that the symbolic system, or language, necessary for communicative action differs fundamentally from that necessary for instrumental action. Instrumental action is monologic: the technical manipulation by a person of some aspect of non-human nature generates an asymmetrical relationship between them rather than a relationship of dialogue. It suffices for such technical action that the human being be able unambiguously to classify the manipulable objects according to precisely definable characteristics; the languages which human beings set up for this purpose allow them to make precise and clear statements which admit of no, or only a very limited degree of, variety of interpretation. Such languages Habermas calls "pure" (1972, p. 161); they are the kind of languages which admit of formalization and pure logical analysis, and were the main object of analysis by the logical positivists, who saw them as the necessary condition for objective science. Communicative action, on the other hand, is dialogic, and its purpose is to allow individuals to communicate about their experience within a shared framework of cultural meanings. The languages in which this takes place must be more flexible than 'pure' languages, for they must allow a mediation between individual experience and shared categories of understanding. But this means that there is always in the statements of ordinary, or natural, language a more or less wide degree of variation of interpretation. The expressions of ordinary language always require interpretation with reference to the context in which they are made; ordinary language cannot be formalized, but remains characterized by "reflexivity" (1972, pp. 162ff.).[17] There is therefore an ever-present need for the 'work' of hermeneutic interpretation, and the cultural sciences which take the domain of communicative action and its products for their objects, must make use of procedures of interpretation which differ only in degree of methodological discipline from the ordinary procedures of mutual understanding in everyday life.

Habermas follows Dilthey in describing these procedures in terms of the 'hermeneutic circle': to understand a cultural object or 'text' involves the interpretation of specific individual meanings within a framework of more general and shared cultural meanings, and this requires a 'circular' or 'back and forth' movement of interpretative thought (see Habermas, 1972, pp. 171f.). There is a mediation between the interpretative framework of the interpreter and the preinterpretations contained in the texts in virtue of their being cultural, and not natural, objects. Unlike the fitting of descriptions of natural objects into classificatory schemes (which can also contain an element of mutual adjustment), this mediation is at the level of communicative action, and involves the interpreter in the practical activity of being able to speak the languages of the texts which he interprets. Unlike the monologic activity of describing natural objects in the categories of a pure language, hermeneutic interpretation is dialogic activity, mediating individually interpreted experience through shared systems of cultural meaning.

By reflection on the procedures of hermeneutic interpretation, Habermas derives the epistemological basis for them in terms of their pragmatic function in the life of the human species. In the case of the methodological procedures of the natural sciences, we saw how Habermas, developing a line of thought in Peirce, showed that those methods gained their meaning from being embedded in the pragmatic context of the instrumental manipulation of objectified nature. In a parallel argument, Habermas shows that the methods of hermeneutic understanding also gain their meaning from being embedded in the pragmatic context of communicative action. Dilthey had already had an inkling of this:

> Understanding first arises in the interest of practical life. Here people are dependent on intercourse with one another. They must make themselves understandable to one another. One must know what the other wants. Thus the elementary forms of understanding come into being (Rickman, 1976, p. 220).

From this, Habermas develops the idea that hermeneutic understanding serves the pragmatic function of creating or restoring necessary mutual understanding:

In its very structure hermeneutic understanding is designed to guarantee, with cultural traditions, the possible action-orienting self-understanding of individuals and groups as well as reciprocal understanding between individuals and groups. It makes possible the form of unconstrained consensus and the type of open intersubjectivity on which communicative action depends. It averts the dangers of communication breakdown in both dimensions: the vertical one of one's own individual life history and the collective tradition to which one belongs, and the horizontal one of mediating between the traditions of different individuals, groups, and cultures. When these communication flows break off and the intersubjectivity of mutual understanding is either rigidified or falls apart, a condition of survival is disturbed, one that is as elementary as the complementary condition of the success of instrumental action: namely the possibility of unconstrainted agreement and non-violent recognition. Because this is the presupposition of practice, *we call the knowledge-constitutive interest of the cultural sciences 'practical'*. It is distinguished from the technical cognitive interest in that it aims not at the comprehension of an objectified reality but at the maintenance of the intersubjectivity of mutual understanding, within whose horizon reality can first appear as something (1972, p. 176).

The practical interest serves the same function of structuring the constitution of an object-world and a range of possible application for the cultural sciences that the technical interest does for the natural sciences.

Once the cognitive interest underlying the methods of hermeneutic interpretation has been discerned, it is possible to raise the question of the limits of applicability of those methods, and in particular, whether they form an adequate basis, alternative to the methods of the natural sciences, on which the social sciences such as sociology could be developed. Habermas' answer to this question is in the negative. There are two components to his denial that the methods of hermeneutics are sufficient for the social sciences. Firstly, whereas the cultural sciences are concerned with the interpretation of systems of cultural meanings, the 'systematic' social sciences must also be concerned with the investigation of relatively stable (even if historically specific) and widespread empirical regularities in social action. Even if hermeneutic intelligibility is to constitute the criterion for deciding that such a regularity is in principle criticizable and alterable, the methodology for studying such regularities with a view to their transformability will of necessity also draw on the methods of the empirical-analytic, nomological sciences.

For this purpose, the "methodological basis of the historical cultural sciences is obviously too narrow" (1972, p. 185). Secondly, Habermas argues that hermeneutics cannot distinguish between conditions in which mutual understanding and intersubjectivity can be furthered and restored by the interpretation of texts within a cultural tradition, that is by hermeneutic interpretation alone, and conditions in which communication has become 'systematically distorted' by rigid structures of domination and force, so that cultural traditions themselves function as ideologies, and emancipation must proceed by the unmasking of such ideologies and by the transformation of the social structures which they support. For this task too hermeneutic methods are insufficient: if the social sciences are to serve the emancipatory interest in reason, they must not be reduced to merely interpretative or *verstehende* sciences (1970a, pp. 287ff.).[18] The "critically mediated knowledge of laws" (*kritisch vermitteltes Gesetzeswissen*) (1972, p. 310), which a critical social science serving the emancipatory cognitive interest would require, would have to be attained by some combination of the methods of both the nomological and historical-hermeneutic sciences, and neither alone would suffice:

> This problem of the interlocking of *empirical-analytic procedures* with *hermeneutics* and the problem of theory formation in the systematic cultural sciences are of central significance for the logic of the social sciences (1972, p. 185).

However aware the *Geisteswissenschaften* were of their underlying cognitive interest, they could not by themselves develop into critical social science.

However, Habermas argues that discussion of the range of application of the hermeneutic methods of the cultural sciences has been restricted by a failure to reflect on the cognitive interest, the practical interest, that is inherent in those methods. To that extent, he suggests that those who have discussed the methods of the cultural sciences and their claim to knowledge have themselves been locked within the 'objectivist illusion' and that their understanding of those procedures has therefore also been positivistic.

Even Dilthey, who according to Habermas was more than anyone in the hermeneutic tradition conscious of the relationship to practical life built into hermeneutic procedures, failed fully to realize that the conditions for success in hermeneutic interpretation were tied to the dialogue relationship between the interpreter and the cultural tradition within which the text to be interpreted was produced. Instead, Dilthey tended to relapse into a psychologistic account of hermeneutic interpretation, in which the interpreter achieves direct access to the psychic state of the producer of the text: a conception of understanding as 'empathy'. In this empathy model of understanding, the preunderstanding on the part of the interpreter, which plays an essential role in fully hermeneutic interpretation, is condemned as a potential disturbing factor in the process of gaining truly objective knowledge of the meaning which an author has incorporated into the text; such disturbing factors must therefore be controlled by making methods of interpretation more 'scientific', less subject to the preconceptions of the interpreter. Like positivist interpretations of knowledge in the natural sciences, Dilthey and others in the historicist-hermeneutic tradition sought a basis of certainty for knowledge in the *Geisteswissenschaften* which would make such knowledge scientific, and thought they had found such a basis in the direct access by the interpreter to the mental states of others.

But, according to Habermas, this conception is just as positivistic in the cultural sciences as it would be in the natural sciences.

Historicism has taken the understanding of meaning, in which mental facts are supposed to be given in direct evidence, and grafted onto it the objectivist illusion of pure theory. It appears as though the interpreter transposes himself into the horizon of the world or language from which a text derives its meaning. But here, too, the facts are first constituted in relation to the standards that establish them. Just as positivist self-understanding does not take into account explicitly the connection between measurement operations and feedback control, so it eliminates from consideration the interpreter's pre-understanding. Hermeneutic knowledge is always mediated through this pre-understanding, which is derived from the interpreter's initial situation. The world of traditional meaning discloses itself to the interpreter only to the extent that his own world becomes clarified at the same time. The subject of understanding establishes communication between both worlds. He comprehends the substantive content of tradition by *applying* tradition to himself and his situation (1972, pp. 309f.).

In failing fully to accept this practical conception of hermeneutics, Dilthey becomes prey to the objectivist presuppositions of positivism, such as its copy theory of truth, and its contemplative concept of knowledge. In a word, "historicism has become the positivism of the cultural sciences" (1972, p. 303).[19]

We are now in a position to see how close a coincidence there is between Habermas' critique of hermeneutics from the standpoint of critical theory and Albert's critique of it from the standpoint of critical rationalism. Both reject as positivistic a conception of the understanding of systems of meaning in terms of direct access to the mental states of others, either through their life expressions or the texts which they produce. Both reject the attempt to transfer a notion of the certainty of sensory experience to the sphere of the understanding of meaning. Both therefore reject the historicist tradition of the *Geisteswissenschaften* and its claims to generate objective knowledge. To that extent there is agreement between them, and Habermas was correct to reject Albert's earlier suggestion that he, Habermas, intended to "criticize empirical-analytic research itself" and that he wished "to play off the methods of understanding against those of explanation" (Adorno *et al.*, 1976, p. 221). But they reach this partial agreement from quite different starting-points. Albert wants to oppose not just a positivistic misunderstanding of hermeneutics, but also any notion that hermeneutics can form the basis of any kind of scientific knowledge, and hence remains committed to the doctrine of the unity of scientific method. Habermas' concern is rather to perceive the practical cognitive interest which underlies hermeneutic procedures, and to reject as positivistic only those interpretations of hermeneutics which fail to reflect this interest and remain instead, like positivist interpretations of the methods of the natural sciences, caught in the objectivist illusion. His argument that the object-domains of the natural and the cultural sciences are constituted by synthetic activity in two different spheres of human action, instrumental and communicative respectively, is continuous with his critique of the doctrine of the unity of scientific method, and accords to each set of methodological rules its proper place.

7.3. Latent Positivism in the Critical Social Sciences

In the third case, the antipositivism of critical theory is directed against 'positivistic' misunderstandings of critical social sciences which, it is claimed, have arisen in the course of such sciences' development and which distort their critical potential into instrumentally applicable theories. We have seen that the main characteristics of a critical social science are two-fold: firstly, that it combines empirical-analytic and hermeneutic procedures into a 'critically mediated knowledge of laws', which is distinguished by the ability to differentiate between 'genuine' invariant regularities and historically specific and humanly transformable regularities on the basis of the latter's hermeneutic understandability or intelligibility; secondly, that it operates through the category of self-reflection in the service of the interest in emancipation, i.e. it should set off a process of reflection in those whose action is constrained by the historically specific forces discovered, and contribute to their liberation from such constraints.

Critical theorists have taken two bodies of 'theoretical practice' as models for such a critical social science: Freudian psychoanalysis and Marxian ideology-criticism. In each case, however, the critical force of the theory has to be rescued from what are claimed to be positivistic misunderstandings which are to be found even in the writings of their founders. To give a flavour of this argument at this point, it will be sufficient to follow Albrecht Wellmer's (1971, Chap. 2) discussion of the nature of Marx's ideology-criticism and the latent positivism of the theory of historical materialism.[20]

In his early writings, Marx developed Hegel's examination of the state produced by the American and French revolutions into a critique of the ideology of the constitutional state. This form of state claimed to realize the rational ideals of freedom and equality before the law. In fact, Hegel saw, this freedom was the freedom of the egotistic individual to pursue his own ends, in competition with all other individuals, in a legal framework which gave all individuals the same formal rights. The 'civil society', whose internal competitive struggle was thereby unleashed, was no longer held together by the ties of traditional morality, and the divisive consequences of economic competition, the intensification of the division of labour and the increase of inequality, had to be compensated for, so Hegel argued, by the state, a power independent of society which could provide the integrative force removed by the loss of traditional morality. The state would reconcile the individual with the universal interest, and thus turn abstract freedom into concrete freedom.

Marx recognized the validity of this rational ideal, but denied that it was actually realized in the constitutional state. All the political revolution that had produced that state had done was to allow individualistic egoism to be pushed to the limits: it was "the perfection of the materialism of civil society" (1975, p. 233). The concrete freedom which the state claimed to guarantee was in fact in contradiction with the formal freedoms enshrined, for example, in the Declaration of the Rights of Man, and the state was incapable of overcoming this contradiction. It could only be overcome by a new historical transformation, in which political emancipation would be followed by human emancipation through the abolition of the materialism of civil society, i.e. through the abolition of private property. In this transformation, the ideal of substantive freedom already claimed by the constitutional state would be realized. Thus Marx was here using a method of ideology-critique, based on the principles of immanent criticism, which should contribute to radical enlightenment and emancipatory practice.

The aim of this immanent critique of ideology was 'the reform of consciousness', in the course of which

> it will become plain that the world has long dreamed of something, of which it needs only to become conscious for it to possess it in reality (Marx, 1975, p. 209).

This final statement was inadequate, however, and shows that without a further step this method of immanent criticism remained only moral criticism. It was also necessary to elaborate the contradiction between the state and civil society into real social conflict of opposed interests, to locate in this conflict an agent capable of and interested in bringing about human emancipation, and to show that the development of existing society necessarily brought about conditions which made social transformation possible.

The ideology-critical method was thus continued in the critique of political economy, but all the more consistently since it also tackles these further problems. The critique of political economy develops the criticism of the bourgeois ideals of equality and freedom, showing that these abstract ideals were merely the "idealized expressions" of the system of economic relationships between individual, self-seeking private producers (Marx, 1973, pp. 244f.). Once it is realized that a developed capitalist economy is a system of exploitative relationships between antagonistic classes, the reality of inequality and unfreedom may be discerned behind the mask of these abstract ideals (1973, p. 249). At the same time, by showing that profit is generated by exploitation in the process of production and not in the course of exchange, the critique of political economy is able to cut through the capitalist legitimating ideology based on the idea of exchange of equivalents (Marx, 1976, pp. 301f.). Finally, the laws of political economy are studied not from the standpoint of their invariance and universality, but from the standpoint of their historical specificity and transformability. Even in the early writings

Marx had seen that "political economy has merely formulated the laws of estranged labour", and had grasped that what was required by a critical science was, in Habermas' words, "a critically mediated knowledge of laws":

> Political economy proceeds from the fact of private property. It does not explain it. It grasps the *material* process of private property, the process through which it actually passes, in general and abstract formulae which it then takes as *laws*. It does not *comprehend* these laws, i.e. it does not show how they arise from the nature of private property (Marx, 1975, p. 322).

But these laws, so Marx argued, are not universal or invariant laws, but "are as such only moments of a historic process" (1973, p. 832). The mistake of political economy, a mistake thrust upon it by the nature of the capitalist mode of production itself, was to take as relations between things what were in fact relations between human beings, and hence to take as invariant what was in fact historically variable. The concepts used by political economists were not perceived by them to be historically specific (1973, p. 832). Such concepts are, however, historical concepts: like the social relationships they describe, they are

> the result of a past historical development, the product of many economic revolutions, of the extinction of a whole series of older formations of social production. The economic categories . . . bear a historical imprint (Marx, 1976, p. 273).

The proper method of a critical approach to political economy, oriented to the transformation of the capitalist mode of production, must involve the examination of the development of such categories; it will then be seen that the general categories, indifferent to specific variations, are themselves the product of the capitalist mode. The critique of political economy must therefore include the critique of the "abstract universals" used by bourgeois economics (1973, pp. 100ff.).

The critique of political economy adds to this purely ideology-critical approach a demonstration that the historical transformation which the critique envisages is not just an empty utopian dream, but one for which the real historically necessary conditions are produced by the development of capitalism itself. The continuous revolution in the forces of production forced on the owners of capital by competition provides the technical conditions necessary for a classless society; the penetration of capitalist relations into the whole world establishes the conditions for a world-wide revolution; the concentration and centralization of capital generate an organized proletariat capable of planning and carrying through a revolution; and the dissolution of traditional relationships and traditional morality provides the conditions for the emergence of a revolutionary consciousness (Marx and Engels, 1968, pp. 46ff.).

However, in order that the Marxian critique may be interpreted as a critical social science in the sense of critical theory, it must be capable in addition of contributing to the process of enlightenment of the social group to which it is addressed, namely the proletariat. Wellmer argues that the ideology-critical method must also be combined with the theory of historical materialism, in order that the idea of a possible future social freedom may be conceptualized in terms of the supersession of a historically developed alienation. By interpreting history as the process of the self-creation of the human species through social labour, the theory of historical materialism provides a materialist interpretation of the abolition of alienation, and thereby also provides

the categories in which the revolutionary transition can be theoretically anticipated as a practical process, and therefore the basis for the conception of revolutionary theory as a theory of revolution (Wellmer, 1971, p. 86).

Through these categories, the proletariat may understand human history and the potential emancipation which it alone is capable of bringing about; critical theory may thus become part of critical practice. But this practice must be a liberating practice which is anticipated even within the capitalist system; it must be a process of self-formative enlightenment, which takes place within the sphere of self-reflection. This has consequences for the proper social relationships within the revolutionary proletariat; if self-reflection is to have emancipatory consequences, the proletariat must act rather than be acted upon, and this requires that its own internal relationships should themselves be based on "undistorted communication":

the connection between enlightenment, practical emancipation, and liberating practice, implies that in the forms of organization and in the principles of the revolutionary movement, certain elements of future social freedom must already be realized, if the revolution is to signify the realization of this social freedom (1971, p. 88).

The proletariat must be able to achieve a collective process of education and self-reflection, to collectively decide upon goals and methods, in order that the process of superseding the rigidities of the old society should not itself generate new forms of rigidity in the new society.

The burden of Wellmer's subsequent argument is that this conception of Marx's ideology-critical approach as a model of critical theory, which seems to do justice to one continuous line of thought from his earlier to his later writings, is contradicted by another, positivistic, interpretation of this potentially critical social science, which was both a component of Marx's own understanding of his method, and a highly influential reading of Marx by later Marxists which has distorted the shape of socialist movements. It is therefore important to expose and criticize this latent positivism, so that the full critical force of the ideology-critical method may be retrieved and further developed.

This latent positivism may be discerned in the theory of ideology which appears in the first systematic outline of the theory of historical materialism, *The German Ideology*. Whereas, as we have seen, in the ideology-critical approach ideology was conceived as "simultaneously illusion ensuring domination and uncomprehended utopia" (Wellmer, 1971, p. 99), so that the ideal of reason embedded in the ideology can be used as a basis of its immanent criticism, in the theory of historical materialism ideology is conceived on the model of the base-superstructure analogy, so that 'the dominant ideas' are merely

> the ideal expression of the dominant material relationships, the dominant material relationships grasped as ideas (Marx and Engels, 1968, p. 61).

As such, ideologies have no autonomous force, but are purely a function of the forms of domination enshrined in modes of productive labour. The history of the human species can be understood as the succession of different modes of productive labour, and the various systems of social relationships, or forms of social intercourse, are conceptualized

> so to speak, only as secondary productive forces, whose function is to make possible the application and development of the primary productive forces (Wellmer, 1971, p. 92).

Systems of social relationships which block the development of new forms of productive labour are, on the other hand, 'fetters' which are necessarily removed and replaced by more conducive relationships. Thus, the 'totality' of modes of productive labour and the systems of social relationships, together with their ideal form, their ideology, can

> be apprehended only as that of a self-regulated dynamic system, which possesses, as it were, only one degree of freedom of historical development: that of technical innovation. Innovations in the sub-system of production give rise to disturbances of equilibrium, which then have to be remedied by adaptation of the subsystem of institutions to that of production. This adaptation, however, is to be comprehended as the 'technically', straightforwardly defined solution of a systemic problem (Wellmer, 1971, p. 93).

To the extent that this can be reconciled with an interpretation of history as a history of class struggles, it can be done only by comprehending the emergence of class conflict as the necessary outcome of such systematic disequilibrium, of a "contradiction between the productive forces and the relations of production" conceived of as an 'objective' contradiction to which there is a technically definable, scientifically discoverable, optimal resolution. The point of the criticism of ideology, in this version, is to make possible the transition from ideological representations of such objective contradictions to 'scientific' understanding of them, which will be carried by the revolutionary proletariat and which will enable systemic disequilibria to be overcome scientifically.

This 'positivistic' aspect of the theory of historical materialism, Wellmer argues, appears in Marx's later writings, both in *Grundrisse* and in *Capital*, in the forms of theories of the necessary transformation of capitalism by reason of its own inherent dynamics, and in 'objectivist' theories of revolution, according to which the revolutionary overthrow of capitalism proceeds according to inexorable historical laws. In the *Grundrisse* Marx appears to represent the progressive application of science to the productive process as if it would lead automatically to the dissolution of the capitalist system. The more the productive process is automated, the less it is dependent on the direct input of human manual labour. The outcome of this process of scientific rationalization of production must be the immanent self-transformation of capitalism.

Any distinction between capitalist and socialist rationalization, with an associated critique of alienated labour under conditions of machinery, becomes blurred, so that socialist rationalization becomes merely an extension of its capitalist origins. The scientific organization of production will still determine and constrain the socially necessary work of regulating and maintaining the machinery, even under socialism, so it becomes unclear that the abolition of capitalism can be understood as a "qualitative change in the sphere of the labour process" (Wellmer, 1971, p. 113). If this is so, Wellmer argues, then the anticipatory force of the self-reflection promoted by ideology-criticism has become redundant; the transformation of capitalism has become historically inevitable and the goals of revolutionary change scientifically specifiable, so there is no need for

> the production of a democratic public attainable only by enlightenment and collective emancipation (1971, p. 113).

This line of argument is reinforced by analysis of the passage in the third volume of *Capital* where Marx argues that the realm of freedom, which the socialist revolution will bring about, lies strictly outside the area of material production (Marx, 1981, pp. 958ff.). The basis for this realm of freedom is the scientific regulation of the realm of necessity, the sphere of material production. In this sphere, there is no room for democratic decision-making of associated producers concerning the relationship between what is technically possible and what practical goals they want to set themselves; rather the organization of material production is that sphere of life which must be technically regulated on the basis of knowledge of scientific laws from which can be derived systemic optimal solutions.

Thus socialist revolution appears not as collective emancipation from historically specific 'frozen relations of dependence' whose constraining force only seems to be the operation of unchangeable laws, but as the institutionalization of the application of scientific rationality based on knowledge of laws that are really unchangeable. Nor is emancipation guided by ideas of the 'good life' which have been rescued from the distorted form in which they appear in ideology; these are replaced by conceptions of technically best solutions to systemic problems of industrial organization.

Wellmer concedes that these implications of the positivistic misunderstanding of Marx's critical theory may well diverge from Marx's own intentions (1971, p. 118). It may well be that, in order to demonstrate that a latent positivism had become actual, Wellmer has produced sharper interpretations of passages in Marx's writings than is merited by their lack of exactness.[21] That positivistic interpretations of Marx have been influential is, however, hardly contentious among those who are not bound by the orthodoxies of Marxism-Leninism, and has been a continuous theme from the early writings of Korsch and Lukács to the critiques of the Soviet Union of the later Frankfurt School (Korsch, 1970; Lukács, 1971; Marcuse, 1971). The specific nature of critical theory's critique of the latent positivism in Marx is to see it as the degeneration of an ideology-critical critical theory into a positivistic theory of revolution precisely because of certain assumptions of the materialist theory of history which were incompatible with the critical theory. From the epistemological point of view, which is the primary concern here, this meant a failure to distinguish sufficiently between the categorial framework of a critical theory, operating in the sphere of self-reflection and guided by the interest in emancipation, and that of an empirical science, operating in the sphere of labour or instrumental action and guided by the interest in technical control. Marx's epistemology, as we have seen earlier, was sufficiently antipositivistic not to be subject to the objectivist illusion as far as the natural sciences were concerned, and his pragmatistic understanding of the natural sciences can form a starting point for an investigation into their technical interest and its relationship to different modes of production. But just because of the essential place of the expansion of the productivity of labour in establishing the material conditions for human emancipation, there is a danger of conceiving human emancipation *as* the expansion of the productivity of labour and the institutionalization of social relationships conducive to that expansion. Thereby an emancipatory critical theory is reduced to a technically applicable quasi-natural science, a positivism latent in the materialist theory of history itself.

CHAPTER 8

THE ANTIPOSITIVISM OF SCIENTIFIC REALISM

Certain difficulties stand in the way of a concise presentation of realism in the philosophy of the social sciences. Only recently have writers on the social sciences become aware of the existence of realism as a specific interpretation of the methods of the natural sciences which might have consequences for the philosophical interpretation of the social sciences; similarly recently have writers propounding a realist view of the natural sciences turned their attention to the social sciences. Unlike critical rationalism and critical theory, scientific realism in the social sciences does not really constitute a 'school'; the former traditions are characterized by a fairly well-defined set of basic assumptions, however varied might be the work of individual adherents, but the implications drawn for the social sciences by those drawn to scientific realism do not cohere into a single framework, and have little to unite them apart from nominal adherence to 'realism'. However, the main lines of the identification and negative critique of positivism are common ground among all 'realists' and form the starting-point of their positive proposals; this common ground will be the basis for the following two sections. In the third section, an outline will be presented of two main developments which have been based on realist principles: the 'ethogenic' social psychology of Harré and his associates, and some attempts to produce a realist interpretation of Marxism.

8.1. Realism's Identification of Positivism

As we have seen in Section 4.2, positivism as a theory of the natural sciences is identified by realism in terms of the basic assumptions and methodological postulates of deductivism. The basic assumptions consist of an ontology of atomistic events, a view that propositions are the only proper vehicle for rational thought, and the belief that the ideal form for knowledge, and especially for scientific knowledge, is the deductive system. From these assumptions all the essential empiricist methodological postulates can be derived: the deductive-nomological model of explanation, the hypothetico-deductive method, the symmetry of explanation and prediction, the principles of empirical invariance and of instance-confirmation or -falsification (Nicod's criterion), and so on.

Essential to the extension of this identification of positivism into the sphere of the social sciences is realism's distinction between 'positivism' and 'naturalism'. In the past, it is claimed, the question of the relationship between the methods of the social and the natural sciences has been discussed in terms of two categorial dichotomies: positivism and idealism, and naturalism and anti-naturalism. These two dichotomies have, however, been treated as if they coincided, so that 'naturalism', or the doctrine that the methods of the social sciences should be the same as those of the natural sciences, has been treated as equivalent to 'positivism', with the consequence that anti-naturalistic doctrines would have to envisage the methods of the social sciences as 'idealist'. Once it has been perceived that positivism is only one interpretation of the natural sciences among others, the two dichotomies can be fruitfully separated, so that it becomes possible to conceive of antipositivist doctrines of naturalism as well as positivist ones. Positivistic naturalism would be the doctrine that the methods of the social sciences should be the same as the methods of the natural sciences, as the latter are described in terms of the basic assumptions and methodological postulates of deductivism/positivism (Keat and Urry, 1975, p. 1; Bhaskar, 1978b, p. 1).

Perhaps it is a sign of changing fashions that it is rare to find the relationship between these two dichotomies, once separated, represented by means of a four-fold table (but see Halfpenny, 1976, p. 32). If the dichotomy 'positivism-idealism' is amended to 'positivism-nonpositivism', the table of relationships shown in Figure 2 may be constructed.

	positivism	non-positivism
naturalism	A	B
antinaturalism	C	D

Fig. 2. Relationships between positivism and naturalism.

When positivism has been identified, for the natural sciences, in terms of the principles of deductivism, interpretations may be given for doctrines which could be allocated to each of the four cells:

(A) 'Positivist naturalism', as already explained, would be the doctrine that the methodological postulates of deductivism are to be applied in the same way in the social sciences as in the natural sciences. From the standpoint of realism, as will be seen in the next section, this doctrine is to be rejected on the same grounds that are used to reject deductivism in the natural sciences.

(B) 'Non-positivist naturalism' would be the term applied to doctrines which embraced a non-positivistic account of the methodology of the natural sciences, and in addition claimed that this same methodology should be used in the social sciences too. As we shall see in Section 8.3, this is the approach taken by scientific realists.

(C) 'Positivist antinaturalism' could be interpreted as a doctrine which accepted a positivist, i.e. deductivist account of the methodology of the natural sciences, but argued that that methodology was not applicable to the social sciences. Its opposition to the doctrine of the unity of scientific method would therefore be based on its positivistic view of the natural sciences.[22] Realists are not the only ones to have critically identified this doctrine; thus Giedymin, from a conventionalist standpoint which he sees as opposed to positivism, writes:

> Antinaturalism in nineteenth and twentieth centuries has been in general due to the acceptance by anti-naturalist philosophers of either the positivist (phenomenalism, physicalism) or of mechanistic-deterministic philosophy of natural science and to their refusal to extend the principles of those philosophical doctrines to the social sciences and humanities (1975, p. 284).

Similarly, Bhaskar, from a realist standpoint, reconstructs traditional debates concerning the relationships between the natural and the social sciences as a dispute between a naturalist tradition based on positivist principles, and a rival antinaturalist tradition which

> has posited, by contrast, a radical distinction in method between the natural and social sciences, flowing from and grounded in the idea of a radical distinction in their subject matters . . . the great error that unites these disputants is their acceptance of an essentially positivist account of natural science (1978b, p. 1).

Doctrines which appear in cell C, positivist antinaturalism, are therefore to be identified as forms of positivism just as much as doctrines which appear in cell A. For realism, it is just as positivistic to reject the unity of scientific method for the wrong reasons as it is to accept it for those wrong reasons.

(D) 'Non-positivist antinaturalism' would be that category of doctrines which rejected a positivist (i.e., in this context, a deductivist) account of the methodology of the natural sciences, and also rejected the doctrine of the unity of scientific method. This category of doctrines is the one least discussed by those, such as realists and conventionalists, who wish to stress the importance of the distinction between positivism and naturalism.[23] The realist writers considered here seem to pay little or no attention to the possibility that the rejection of a deductivist account of the natural sciences should be thought to have no particular consequences for the methodology of the social sciences, but should rather leave intact the arguments for rejecting any doctrine of the unity of scientific method. It will be argued, in Chapter 9, that realists have at present no convincing way of tackling this problem.

In summary, then, we may say that realism identifies as forms of positivism all those doctrines which can be allocated to cells A and C of Figure 2, namely both naturalist and antinaturalist doctrines which incorporate a deductivist account of the methodology of the natural sciences.

8.2. Realism's Critique of Positivism

It has been shown, in Section 4.3, that realism's critique of positivism in the theory of the natural sciences rested on two principles of criticism. The first principle is that the basic assumptions of deductivism generate problems that are insoluble within the terms of positivism; it was suggested that this form of criticism is necessarily inconclusive, but is made much more persuasive by the ability of realism to restate those problems according to its own assumptions in such a way that solutions may be advanced or the problem made to disappear. The negative critique is thus reinforced by a method of confrontation of positivist and realist theories. The second principle of criticism is based on the attempt to show that the assumptions of positivism generate descriptions of scientific practice in the natural sciences which are incorrect or partial, and which therefore erect false ideals for scientific activity and reasoning. Without being able to examine the detailed evidence and argument advanced on the basis of this principle, it was nonetheless decided to accept this claim as plausible for the purposes of the present work.

Realism's critique of positivism in the theory of the social sciences is essentially a transference of the critique of positivism in the theory of the natural sciences into the assessment of doctrines of positivist naturalism and positivist antinaturalism. In the process of this transference, the second principle of criticism summarized in the previous paragraph takes on a different significance. Whereas in the case of the natural sciences, the realist rejects positivism because it fails to do justice to 'the persistent intuitions of scientists' as embodied in their scientific practice, in the case of the social sciences realism is concerned less to do justice to any persistent intuitions which social scientists may incorporate in their practice than to show how social scientists have been misled by positivist accounts of the natural sciences into the blind alleys of both positivist naturalism and positivist antinaturalism. The reconstruction of actual scientific practice in the natural sciences thus takes on normative force which must exercise its influence on the social sciences too. This is understandable in the light of the conclusion reached in Chapter 5, that scientific realism, trapped within the objectivist illusion, believes itself to have discovered the essence of science, to have described the only possible methodology which is in accord with the real nature of the world. Despite this objectivism, whose consequences for realist theories of the social sciences will later be examined, there could nonetheless be considerable substance in the central idea of the realist critique, that social scientists have adopted both naturalist and antinaturalist doctrines on the basis of an undifferentiatedly positivist conception of the methodology of the natural sciences. It will therefore be useful to outline some examples of this form of argument that have been presented by realist writers.

8.2.1. Realism's Critique of Positivist Naturalism

We may first outline the realist critiques of positivist naturalism. Harré's own extension of his critique of positivism into the sphere of the social sciences has been primarily concerned not with sociology but with social psychology. Together with P. F. Secord, Harré produced in *The Explanation of Social Behaviour* (1972, pp. 29ff.) an outline of a realist social psychology which is prefaced by a critique of what they see as the misguided positivist naturalism prevailing in contemporary social psychology. Social psychologists, like psychologists in general, have believed that the possibility of developing social psychology as a science depends on following essentially the same methodological postulates as those which they assume to be the basis of the natural sciences. In pursuance of that belief, Harré and Secord argue, they took over a conception of scientific methodology drawn from the most basic form of positivism, logical positivism or logical empiricism. They therefore accepted the atomistic ontology and the atomistic sensationalist epistemology which formed the basis of that extreme empiricism. They took over operationism as a theory of concept-formation, and verificationism as a theory of meaning. They were thus constrained to a Humean, regularity theory of causation, and the regularity-determinism associated with it. In addition they adopted a 'mechanistic model of man'. The most consistent outcome of this positivist naturalism was the behaviourism of Watson and Hull. The atomism of positivism here manifested itself as a search for simple units of behaviour, whose relationship to simple stimuli could be studied by laboratory experiment and summarized in causal propositions interpreted as regularities of sequence. More complex, social behaviour would eventually be understood in terms of the combination of laws explaining simple behaviour. Even though Skinnerian 'radical behaviourism' claims to reject logical positivism, it is still, Harré and Secord argue, tied to the mechanistic and Humean conceptions derived from a positivist conception of the natural sciences (1972, pp. 34f.).

Criticism of stimulus-response models of human behaviour and of behaviourism in general is of course by no means specific to scientific realism, nor is criticism of its narrowly positivist assumptions (see Taylor, 1964). In fact, Harré and Secord neither claim such specificity for their views (1972, p. 2), nor do they even present a detailed critique of what they see as dominant behaviourism.

The main lines of criticism, which are sketched out lightly prior to the elaboration of their own approach, depend mainly on behaviourism's alleged inability to do justice to what is both specific and essential to human beings as opposed to animals, namely their ability to actively interpret the situation in which they find themselves in terms of a shared symbolic system, and to act purposively on the basis of their interpretation of situations, orienting their action to shared rules of appropriate action. From the point of view of this 'anthropomorphic model of man', Harré and Secord criticize the experimental methodology of contemporary social psychology as incapable of studying action in terms of the agents' interpretations, and opposes the experimental abstraction of action from its normal social and cultural setting (1972, pp. 36ff., 48ff.). These criticisms are in no way specific to realism, and are for example almost identical to those advanced by symbolic interactionism (cf. Lindesmith and Strauss, 1968; Manis and Meltzer, 1972). What is specific to realism's critique is that behaviourism's positivist naturalism was in any case derived from an incorrect interpretation of the natural sciences; the assumption seems to be that if psychology had not attempted to imitate something that did not exist, these mistakes would not have been made. Whereas the symbolic interactionist critique therefore operates at the level of substantive theory, the realist critique, although apparently identical in content, is intended to operate at the level of methodology. Despite Harré and Secord's general note of approval for Meadian symbolic interactionism (1972, pp. 41f.), they ought to have extended their critique of positivism to include many manifestations of that tendency as well. For it is apparent that much symbolic interactionism is guided by principles of scientific methodology which do not differ markedly from those followed by behaviourism, however different the substantive theories developed by these two schools might be (cf. Halfpenny, 1976, pp. 219ff.). Thus when Lindesmith and Strauss state that

> the development of a genuine scientific system in the modern sense requires that the implications of the basic position be explored and translated into more specific propositions or into lower-order hypotheses. The acid test of broad encompassing theory then consists of deducing more specific propositions that are suitable as guide lines in research activity and that can be tested by confronting them with the empirical data of human behaviour (1968, p. 11),

they appear to be endorsing principles of deductivism which the realist would reject as positivistic.[24] By restricting their critique to varieties of behaviourism which could quite easily be interpreted as straw man-ism, Harré and Secord thus fail to bring out the full breadth of the realist critique of positivist naturalism.

In contrast, Keat and Urry are less restrictive in their consideration of positivist naturalism in sociology. They expressly concede that neither Durkheim nor his positivist predecessors hold to an extreme positivist position; which would be, in the study of human beings, both methodologically individualistic and behaviourist (1975, p. 80).

Strict positivism, they argue, incorporates a tendency to theoretical reductionism, and the insistence of naturalist sociologists, or at least some of them, that the social realm cannot be reduced to the psychological and hence to the biological and physical therefore comes into tension with their positivism. Keat and Urry are thus concerned not only to criticize reductionist behaviourism, but rather to identify and subject to realist criticism any positivist elements in naturalistic thought about the social sciences, even where such elements are combined with strands from other traditions.

The cases of Comte and Mill are subsumed without difficulty under the critique of positivist naturalism. Both accepted an essentially positivist account of the natural sciences and thought that such positivist methods should be extended to the social sciences. Comte, however, believed in the existence of an independent social realm, irreducible to the realms of sciences lower in the hierarchy. The social realm is governed by laws, of social statics and social dynamics, specific to the social and irreducible to the laws of other sciences. The investigation of such laws, particularly the laws of social dynamics, required special methods, namely the historical method of comparing successive states of human development, in order to discover laws of the inevitable transition from one state to another. From a realist standpoint this conception remains positivist, since social laws are still understood as statements of regular succession between observable events, rather than as attempts to uncover underlying generative mechanisms which produce observable regularities but which are not identical to them. However, Keat and Urry judge Comte to have been inconsistent with his own positivist principles when he uses 'society' as a concept referring to unobservable conditions which are responsible for observable regularities; there is in fact, they argue, considerable unclarity in Comte as to what counts as 'observable' and hence as unobservable (1975, pp. 71ff.). John Stuart Mill was in this respect more consistently positivist than Comte, refusing to accept that there could be knowledge of any hidden causes of phenomena, restricting the concept of observability strictly to what was accessible to sense-experience, modelling the laws of social science on Newtonian mechanics, advocating a strict, methodological individualism as the outcome of an atomistic view of the universe, and proposing the reducibility of the social to the psychological (Keat and Urry, 1975, pp. 75f.).

The case of Herbert Spencer is, according to Keat and Urry, more problematical. Although Spencer may be said to have accepted an essentially positivist view of the natural sciences, he disclaimed the influence of Comte and the positivist movement. Despite this, there are significant similarities in their conceptions of social science, in particular in their rejection of reductionism. Spencer believed that certain sciences, among them sociology, were concerned with totalities of phenomena rather than with individual elements. Another such science was biology, and the organic analogy which he believed essential to social science operated at a methodological rather than a substantive theoretical level, in that the analogy directs attention to similarities in the growth and functional differentiation of animal and social organisms. Despite this 'holism' or 'emergentism', however, Keat and Urry claim that Spencer's use of the organic analogy was, judged from the standpoint of realism, still positivist.

> We have to locate his employment of the organismic source of the model of society within his general positivist desire to identify and establish the law-like regularities of social life existent between empirically observable phenomena. Thus we see that the source is not chosen by Spencer because it will provide us with the means of describing underlying mechanisms and processes productive of observable regularities. Rather we use the organismic source to produce a model of society which comprises a set of law-like relations between observables (1975, p. 79).

Similarly, Spencer's analysis of evolution is as an inductively established law of the increasing differentiation and increasing integration of all kinds of systems, rather than as a guide to the investigation of underlying causal mechanisms responsible for these observable processes at different levels of reality.

The critique of Durkheim's positivist naturalism is more complicated, because non-positivist elements are more clearly identifiable in his writings. Durkheim, as a 'scientific rationalist' (1964b, p. xxxix), wished to extend the methodology of the natural sciences to the study of society, and he conceived of science as the establishment of laws, or relations of cause and effect. He also wished to establish the autonomy of sociology as a science whose laws were irreducible to those of any other science, and was thus concerned to demarcate an object-realm for such an autonomous sociology.

This object-realm was the category of 'social facts', distinguished by their 'externality' to the individual, their generality independent of individual manifestations, and the constraint they impose on individual action. Social facts, like all other kinds of thing, are to be observed by sense-experience, since this is the only source of knowledge. In any particular sociological study, the social facts selected for study should be defined in advance by a set of common external characteristics, which constitute the inherent properties or essential nature of a type of phenomena (1964b, p. 35; cf. Keat and Urry, 1975, p. 84). The cause of this type of social fact must then be sought in other, antecedent social facts, and the results of the investigation expressed in causal laws.

However, despite the emphasis on sense-observability and on the establishment of causal relationships between observable social facts, Durkheim also introduced ideas which do not cohere easily with these positivist notions. In *The Division of Labour in Society*, he argued that

> social solidarity is a completely moral phenomenon which, taken by itself, does not lend itself to exact observation nor indeed to measurement. To proceed to this classification and this comparison, we must substitute for this internal fact which escapes us an external index which symbolizes it and study the former in the light of the latter (1964a, p. 64).

The index for different types of social solidarity was taken to be different types of legal code. More important than the substantive example is the methodological distinction between internal and external facts. The use of the latter as indicators of the former is a reflection of a conception of science which, it is suggested, departs from his otherwise positivist position.

> Science goes from without, from the external and immediately sensible manifestations, to the interior characteristics of which these manifestations betray the existence (Durkheim, 1974, p. 19).

This apparently realist conception of science might be supported by Durkheim's comment on his index of social solidarity:

> we can know causes scientifically only by the effects that they produce, and in order to determine their nature, science chooses from these effects only the most objective and most easily measurable. Science studies heat through the variations in volume which changes in temperature produced in bodies, electricity through its physico-chemical effects, force through movement. Why should social solidarity be an exception? (1964a, p. 66).

But realist writers differ on the interpretation of such statements. Benton sees them as showing that Durkheim

> does adopt realist forms of explanation, and even recognises this in places when he reflects on his own research practice . . . To this extent, Durkheim's conception of scientific knowledge breaks from positivism (1977, p. 87).

Bhaskar concurs with this interpretation (1979, p. 167). Keat and Urry, on other hand, are not so sure. Such passages do show Durkheim reflecting on the status of theoretical terms in science, but

> the relationship between the index and the theoretical term, for example, between repressive law and mechanical solidarity, is unclear and unresolved. Is the former the means by which we confirm the existence and causal efficacy of the given theoretical entity? Or is it the means by which, positivistically, we reduce such entities to empirically meaningful and testable statements in the observation-language? (1975, p. 86.)

They conclude that "Durkheim's position is ambiguous between different conceptions of science".[25] They also suggest that his theory of definition, which involves "classing together all phenomena that happen to share certain common external characteristics", betrays a third conception of science, "essentialism" (1975, p. 85).[26] It remains unclear therefore whether Durkheim partially escapes the realist critique of positivist naturalism or not.

Whatever might have been Durkheim's own understanding of his theory of science, Keat and Urry claim that the way he has been interpreted in later American sociology has been almost exclusively positivist. For example, Durkheim's theory of suicide has been reconstructed as a system of deductively related higher and lower level laws, whose theoretical concepts, such as the degree of social integration, have been defined operationally (instead of being conceived of as attempts to describe underlying generative mechanisms), and whose lowest level deductive consequences, statistical correlations between suicide rates and other operationalized measures, provide the only empirical test of the theory (Keat and Urry, 1975, pp. 82ff.). It is then briefly suggested that this positivist naturalism is typical of much American sociology in general, and the critique of positivist naturalism is thus extended, in rather cavalier fashion, to Parsons, role theory, and studies of stratification (1975, pp. 91ff.).

From these few examples of realist argumentation, the main lines of the realist critique of positivist naturalism are clear. Naturalism as such is not brought under discussion, and the critique does not in general depend on any conception of a difference between the natural and the social realms. Rather, positivist naturalism is condemned for having mistakenly adopted from a misunderstanding of the natural sciences a conception of scientific explanation based on the deductivist principles of positivism. From this, adherents of positivist naturalism are led to forms of research procedure designed to discover regularities in sequences of observable events, the systematization of which into a deductive system is taken to be the construction of scientific theory. From the realist point of view, these research procedures are not so much misguided as impoverished. The attempt to find patterns in observable events is, according to the realist, essential, but is only the starting-point for scientific discovery. As Harré has put it:

> Of course we must try to find correlations between types of occasions and types of social action ... But correlations are not science ... Correlations in nonpositivistic science are the occasions for the investigation of the causal mechanisms responsible for those correlations, the discovery of which would be a contribution to science. Correlational studies remain at the level of natural history (1977a, p. 333).

Social science guided by the tenets of positivist naturalism thus remains at the stage of "critical description" (Harré and Secord, 1972, pp. 68ff.); the advance to scientific explanation would require it to go beyond the surface regularities to the investigation, with the help of models, of the causal generative mechanisms which underlie them.

8.2.2. Realism's Critique of Positivist Antinaturalism

The basic principle of realism's critique of positivist antinaturalism was stated by Russell Keat:

> several anti-naturalist arguments assume a positivist characterization of the natural sciences, and must therefore be reassessed if we take seriously the attacks on positivism (1971, p. 9).

To the extent that adherents of antinaturalism base their opposition to the doctrine of the unity of scientific method on a mistakenly positivist conception of the natural sciences, their arguments must be rejected; and to the extent that antinaturalism is supported by such mistaken arguments, it too must be rejected. Keat did recognize, however, that this kind of negative critique of antinaturalism does not also constitute in itself an argument in favour of naturalism.

One prime target for the realist critique of positivist antinaturalism is Peter Winch's book, *The Idea of a Social Science*, and the arguments advanced in connection with realist criticism of Winch may be taken as typical. Winch, as Keat points out, in effect advertises that his antinaturalism is grounded in a positivist conception of science by his choice of Mill to represent the naturalist opposition. Winch does, it is true, describe Mill's statement of his position as "naive" and refers to "some of the more obvious defects" which later writers have tried to remedy; but it appears that it is Mill's statement of the principles of a naturalist social science which Winch holds to be naive and defective, not the view of the natural sciences on which it is erected (1958, p. 66, but cf. p. 80). Thus Keat is able to suggest that, at a number of points at which Winch refuses to allow what he takes to be the methods of the natural sciences to be applicable to the study of societies, his arguments are in fact antipositivist ones which could apply equally to positivist theories of natural science. When Winch agrees with Wittgenstein that there are cases in which

> we might well be able to make predictions of great accuracy . . . and still not be able to claim any real understanding of what those people were doing (1958, p. 115),

Keat can point out that precisely the same is true in the natural sciences, a state of affairs which leads realists to reject the doctrine of the symmetry of explanation and prediction which both positivists in general and Winch in particular accept as a fundamental component of scientific method (Keat, 1971, p. 10; cf. Winch, 1958, pp. 91ff.). When Winch argues that

> historical explanation is not the application of generalizations and theories to particular instances: it is the tracing of internal relations (1958, p. 133),

Keat can suggest that this rejection of the deductive-nomological model of explanation is equally part of the realist's critique of positivism in the natural sciences, although he concedes that the "internal relations" which Winch has in mind are conceptual relations between rules and rule-following, not those between components of generative causal mechanisms (1971, pp. 10f.). When Winch argues, against Popper's view that it is mistaken to think that scientific models represent some "permanent ghost or essence" lying "within or behind the changing observable events" (Popper, 1961, p. 136), that if this means that models are merely "formulated by the investigator in order to explain certain experiences" then this cannot apply to models of social institutions, which are not "just explanatory models introduced by the social scientist for his own purposes" but rather "govern the way the members of the societies studied by the social scientist behave" (Winch, 1958, p. 127), Keat can reply that this is exactly what a realist would say about models in the natural sciences too; models are not just heuristic devices, as the positivist would have it, but attempts to represent hypothesized generative mechanisms which may turn out to be real (1971, pp. 11f.). Similarly, when Winch argues that understanding a social institution is quite different from establishing a regularity in sequences of observable behaviour (1958, pp. 83ff.), the realist can argue that the scientific discovery of an underlying generative mechanism is also quite different from the establishment of a regularity in sequences of observable events.

Many antinaturalist arguments depend on making a sharp distinction between human action and physical behaviour. Winch's version of the use of this distinction to draw antinaturalist conclusions is typical. Action is meaningful, and gains its meaning from the socially shared rules which identify an action as the kind of action it is. Actions are internally related to the systems of rules which give them meaning, thus their character as a particular kind of action is intrinsic to them; for example, an act of obedience is internally related to the order that preceded it, and the recognition of this relationship establishes the act as intrinsically an act of obedience (1958, pp. 124f.). The rules shared in a society establish standards of appropriate action, and action is to be explained by reference to those standards, which provide actors with reasons for action which both explain and justify (1958, pp. 81f.). All these features differentiate action from physical movement, which is not meaningful, has no intrinsic character, is identified by sense-observation rather than by internal relationship to rules, and is explained causally. From this fundamental distinction the antinaturalist conclusion is drawn that explanation of action in terms of reasons cannot be a form of causal explanation, that action cannot be explained causally at all, and that therefore the methods of the natural sciences are quite inappropriate to the study of social action.[27]

Keat argues that this argument too rests on a positivist understanding of the methods of the natural sciences. The argument that action of a particular type has an intrinsic character while physical movement does not is grounded in the atomism and nominalism which are basic principles of positivism. Thus, taking Winch's own example of a physical event, a clap of thunder, Keat argues that the identification of an event as a clap of thunder is not purely a matter of the atomistic sense-experience of a noise, but rather is dependent on the understanding of the causal relationship of the noise to the electrical storm of which the noise is a component. Knowledge of the causal mechanism responsible for the noise is intrinsic to identification of it as a clap of thunder (Keat, 1971, pp. 13f.; cf. Keat and Urry, 1975, pp. 157f.). Thus causal explanation of a physical event is not based on its regular association with another, independently identifiable, type of event, but depends on being able to describe the causal mechanism which underlies it, and whose intrinsic powers are capable of producing events of that type. Because he assumes a Humean conception of causality, Winch is unable to see the similarity between his description of explaining an action in terms of reasons and a non-Humean, realist, account of causal explanation. On a realist account, the conceptual relationship between reason and action which Winch expresses is no obstacle to the interpretation of that relationship as a causal one (Keat and Urry, 1975, p. 158).

Essentially the same type of critique is directed against, for example, Alfred Schutz's moderate form of antinaturalism based on the inadequacy of sensory observation for understanding the social meaning of action (Keat, 1971, pp. 12f.), and Weber's distinction between direct and explanatory understanding (*aktuelles* and *erklärendes Verstehen*) (Keat and Urry, 1975, pp. 148ff., 158). The outline given of the realist critique of the positivist basis to Winch's antinaturalism is, however, perhaps sufficient to show how realists can extend their antipositivism to cover not just what they take to be misguided forms of naturalism, but also misguided forms of antinaturalism.[28] With that, we may pass to an account, of necessity brief and highly selective, of realists' more positive proposals.

8.3. Beyond Positivism: Realism and Social Science

It has already been explained in Chapter 4 that realism was selected as one of the types of antipositivism to be discussed because of the claims made for it:

that it could constitute the basis for a new and revitalized theory of the unity of science, once the positivist versions of the unity of science had been shown to be bankrupt. There exists in realism, therefore, an inherent prejudice in favour of the development of an antipositivist naturalism, and the possibility that realism's critique of positivism in the theory of the natural sciences could be quite without effect on reflection in the social sciences is from the outset almost removed from consideration. As we suggested above, this is the result of the realists' belief that they have unearthed the 'essence' of science, the only form of scientific knowledge that is capable of capturing the nature of reality. Thus Bhaskar poses the problem of the possibility of social scientific knowledge as follows:

> we have in science a three-phase schema of development, in which in a continuing dialectic, science identifies a phenomenon (or range of phenomena), constructs explanations for it and empirically tests its explanations, leading to the identification of the generative mechanism at work, which now becomes the phenomenon to be explained, and so on. On this view of science its *essence* lies in the move at any one level from manifest phenomena to the structures that generate them. The question of naturalism can thus be posed as follows: to what extent is it possible to suppose that a comparable move can be made in the domain of the social sciences? (1978b, p. 4; emphasis supplied).

The assumption is that only to the extent that those who study society and social action can also follow the realist methods of the natural sciences can the knowledge they produce claim to be scientific.

This does not mean that the methods of research in the social sciences should be identical to those in the natural sciences. Realist naturalism must hold

> that it is possible to give an account of science under which the proper and more or less specific methods of both the natural and social sciences can fall. But it does not deny that there are important differences in these methods, grounded in the real differences that exist in their subject-matters (Bhaskar, 1978b, p. 2).

The realist theory of science thus provides a general characterization of the essential features of any science, which minimally include the investigation, with the help of iconic models, of the real causal generative mechanisms responsible for observable phenomena, and then within this general characterization outlines the more specific methods appropriate to the study of different domains of reality. We may expect from these more specific accounts more detailed representations of what kind of generative mechanisms might be sought in particular domains of reality, what kind of iconic models and what sources for such models might be appropriate, what kind of stratified structures might be found in particular domains, and what kind of powers and natures the entities and structures responsible for phenomena might have. These are the issues which should be addressed by a realist naturalism in the context of the theory of the social sciences. With a view to finding out how these issues have been addressed, we may briefly examine two tendencies which have developed out of realist naturalism: the 'ethogenic' social psychology propounded by Harré and Secord and their associates, and some attempts to outline a realist sociology based on a realist interpretation of the work of Marx.

8.3.1. The Realism of Ethogeny

'Ethogeny' is the name given by Harré and Secord to the projected social psychology which would overcome the failings of 'positivistic' social psychology. Positivistic social psychology has, it is claimed, not progressed beyond the stage of critical description: but if it is to be considered a science, social psychology will have to go beyond "the identification of patterns of overt behaviour, and the conditions under which they occur".

> As we have described, the established sciences are distinguished by the fact that they always seek to answer the question as to *why* the observed patterns among phenomena obtain. And they always proceed to answer that question by discovering or imagining the causal mechanisms which produce the phenomena ... The study of that which in the phenomena of social interaction corresponds to causal mechanisms in physical science we have called *ethogeny*. (Harré and Secord, 1972, pp. 127f.; all undated page references in this section refer to this book.)

The use of the word 'corresponds' here is significant. Harré and Secord do not, by and large, represent that which underlies and generates overt social behaviour as *being* causal mechanisms, but as being *analogous* to the causal mechanisms discovered by physical scientists.

The implications and strength of the analogy will be the subject of later discussion; for the moment we may merely summarize Harré and Secord's view of the consequences of the analogy. As we mentioned above, Harré and Secord counterpose to the positivistic 'mechanistic' model of man an alternative 'anthropomorphic' model of man. The main features of this model, common to many other approaches including symbolic interactionism, attribute to human beings not just the liability to react passively to external stimuli but more importantly a set of active powers or capacities. Human beings have the capacity to initiate action and to monitor the performances they initiate. In addition they have the capacity to 'monitor their monitorings' (pp. 89ff.), that is, they can be self-consciously aware of their own monitored performances, so that they can both plan performances in advance of carrying them out, and provide commentaries or accounts of their performances during and after their execution. These capacities are based on, or go together with, the capacity to use language, to represent in a shared symbolic system both to themselves and to others interpretations of the situations in which they must act. Their actions are therefore dependent on their interpretations of situations, and not directly on the 'objective' nature of the stimuli to which they may be exposed. Action is planned on the basis of the meaning of the situation, including the interpretation of the identities of the agents in a situation, for human beings are not just biological individuals, but also take on social identities, or personas, which need not be unitary for any biological individual. Human beings have the capacity, derived from their capacity to monitor their performances, to adopt a variety of different social identities, or selves, and present these more or less consistently for the interpretation of others. They also orient their action to systems of rules of appropriate behaviour, which vary for occupants of different social identities, and justify their action, if necessary, by commentaries which make reference to those rules (pp. 87ff.).[29]

From these two models of man two different models for the explanation of the pattern of behaviour observed in a social 'episode' may be derived. The mechanistic model provides a schema for the explanation of 'causal episodes', in which the sequences of events are related by the operation of a causal mechanism, and human beings react passively to the push and pull of forces exerted by the environment. The anthropomorphic model, on the other hand, provides a schema for the explanation of 'formal episodes' (pp. 168f., 179ff.); patterns of observable action are seen as being generated by the conscious following by self-conscious self-monitoring agents of an explicit set of rules which apply to a given type of situation. A typical formal episode is a marriage ceremony; the sequence of actions observed in such a formal ceremony are adequately explained, according to Harré and Secord, by reference to the explicit rules which state what each person should do in the performance of the ceremony. It is also simple to discover "the ethogeny of a formal episode like a marriage ceremony", since the direct accounts of the participants of their performances will refer to the explicit rules which they followed (p. 129). Unlike the sequence of actions in a causal episode, in the case of a formal episode

> component actions within the sequence do not in general lead to later members of the sequence through the causal action of biological mechanisms, but are related through their meanings, and the following of rules. For instance, the sequence of actions which several people perform which lead to two people coming to be married constitute a formal episode, whose principles of order are in fact explicit rules as to what must be done by the occupant of each role and in what order. As with many formal episodes the point of following the rules is to do the actions pursuant upon the performance of an act (pp. 168f.).

Thus the conscious self-monitoring following of action-generating rules corresponds, for formal episodes, to the causal generative mechanisms for all causal episodes, including those studied by physical scientists; and the discovery of the former is the basis for the development of a scientific social psychology (p. 142; cf. Harré, 1974).

However, most of the episodes of social life can be allocated unambiguously neither to the category of causal episodes nor to that of formal episodes: they are rather, Harré and Secord claim, "enigmatic episodes" (p. 171). Enigmatic episodes are those which cannot be immediately explained either as generated by causal mechanisms whose working is well known or as generated by explicit rules or well-established conventions. If social psychology is to penetrate beneath the observable sequences of action of enigmatic episodes, it has available to it only two schemata of explanation, those appropriate to causal and formal episodes respectively. Harré and Secord argue that traditional, positivistic social psychology, to the extent that it has gone beyond critical description, has attempted to assimilate enigmatic episodes to causal ones, and to explain them using the categories of causal explanation. The ethogenic approach is based on the proposal to attempt to assimilate enigmatic episodes to formal episodes, to treat the sequences of action in them *as if* they had been generated by the following of rules or conventions, even though the rules are not explicit and the following of them is not consciously monitored by the agents. The method of the imaginative construction of models of rule-following which could have generated an observable pattern of behaviour is seen as analogous to the imaginative construction in the physical sciences of models of causal mechanisms which could have generated patterns of observable physical events. In each case, the hypothetical model should be subject to empirical testing: its 'reality' should be checkable against empirical evidence.

The structure of formal episodes is thus used as the source of models for the unknown structure of enigmatic episodes. For this purpose, Harré and Secord develop a classification of formal episodes which can be used as models for the description and explanation of enigmatic ones. A formal episode can be analyzed as a situation in which a conventional set of roles are available to be filled, and a set of rules specify the behaviour appropriate for the occupant of each role; formal episodes thus provide 'rule-role models'.

Such rule-role models may be classified on various dimensions, and typical models may be described. Among the models discussed by Harré and Secord are 'liturgies', in which a 'script' is specified for a ceremony, rules are laid down prescribing what actions are to be performed and how, and role-conditions specify which people are qualified to perform different parts of the script; 'games', in which action is directed towards an outcome sought by the participants, governed by rules of play, and subject to principles of skill and the application of strategy; 'routines', in which sequences of action are generated by the following of rules, but in which, unlike the case of ceremonies, the outcome of the action is determined causally rather than conventionally; and 'entertainments', in which a sequence of action is not directed to a specific outcome, either causal or conventional (pp. 188–205). These different types of formal episodes are available to act as models for enigmatic episodes, which can be described, more or less metaphorically, in terms of the concepts drawn from a particular type of formal episode, and an underlying rule-role structure hypothesized.

In order that a model drawn from a formal episode may be applied to an enigmatic one, a way of interpreting a sequence of interaction must be available which goes beyond the natural attitude of the unself-conscious participant. Harré and Secord argue that the adoption of 'the dramaturgical standpoint' serves this purpose. It is explicitly claimed that the analysis of agents' accounts of their performances in episodes, by an analyst who takes the dramaturgical standpoint, is analogous to the use of microscopes (and by implication any other of the material forces of inquiry) in the physical sciences (p. 205). The procedure is described as follows:

In taking the dramaturgical standpoint we ask ourselves how we would perceive what we are doing were we acting deliberately, and following a script as in a play. By taking the dramaturgical standpoint we are in a position to discover not only the general nature of what we are doing, that is, which of the general types of Rule-role model is appropriate to the analysis of the episode, but under the guidance of that discovery to produce a commentary detailing the rules, plans and conventions we are following and the meanings which the several parts of the performance have for us. Thus adopting the dramaturgical standpoint is a way of attaining the general position from which it becomes possible to discover the ethogeny of an episode, that is the rules, meanings, and so on which are followed by a conscious agent in doing and saying the things that make up the overt structure of the episode. Thinking of the performance as if it were a play demands that we pay attention to just those features we have drawn attention to ... as the system of reasons for the actions which constitute the monitoring, anticipatory or retrospective commentary in a justificatory context. This standpoint can be taken vicariously by an onlooker as well as actually by a performer (p. 208).

In other words, the dramaturgical standpoint enables either a participant or an observer to reinterpret an episode whose structure is neither obviously formal nor obviously causal in terms of the categories of a type of formal episode, by imagining the episode as if it were a game, a ceremony, or whatever, consciously being played out by the participants. The accounts, or commentaries, given by agents in the situation can also by this imaginative adoption of a model be reinterpreted in terms which are drawn from the model. Thus imagining a sequence of interaction as if it were a 'character contest', a species of 'moral game', to take an example from the work of Erving Goffman, is treated by Harré and Secord as analogous to imagining an atom as if it were like a planetary system on a minute scale (pp. 224f.; cf. Goffman, 1969, pp. 181ff.).

The problem faced by this procedure of imaginative redescription, both in the physical sciences and in ethogenic social psychology, is, according to Harré and Secord, also analogous. In the physical sciences a hypothesized model of a generative causal mechanism typically postulates the existence of unobservable entities; these existential hypotheses become testable when scientific instruments become available which make it possible, however indirectly, to observe such entities through their observable effects. In this way a check is possible on the scientific imagination through which the model was constructed. The same problem exists in ethogeny; as Harré and Secord put it:

> How do we know when it is appropriate to apply one of these models to an episode, or to use a conceptual system in the description of an Act-action structure that may be some suitable combination of concepts derived from the group as a whole? And how do we know when this has been done well or ill, correctly or incorrectly? (pp. 229f.)

In other words, the difficulty exists of finding a procedure that corresponds to the testing of an existential hypothesis, that provides an empirical check on the imagination of models of episodes. Harré and Secord's answer to this problem draws on what they (and, it may be observed, critical theorists too) see as an essential difference between the physical and the social sciences. They recognize that the relationship between the social psychologist and the people whose action is being interpreted is quite different from the relationship between a physical scientist and natural objects. Social action, it is argued, is performed for an audience, and the work of a social psychologist in the interpretation of social interaction places him too in the position of an audience.

> A social psychologist, by adopting one or more of these models, automatically creates an audience by becoming an engaged spectator rather than a detached observer ... It is very important to notice that the audience and one or more of the players may be the same people (pp. 230f.).

In producing an interpretation of a sequence of interaction in terms drawn from a model of a formal episode, a social psychologist is exercising precisely the same capacity for the self-conscious monitoring of action that any agent possesses, and the interpretation produced has the same logical status as any other interpretation generated by a self-monitoring participant. The social psychologist's and the participant's accounts of the structure of an episode are therefore rival accounts of the same episodes from different standpoints, a state of affairs that has no counterpart in the physical sciences.

The empirical checking of the application of a Rule-role model to an episode thus takes place in the context of communication between the social psychologist and the participants. This is true whether the social psychologist is himself a participant actor in the episode, or stands outside it as an observer. In either case, Harré and Secord argue, the empirical checking of accounts of episodes takes the form of *negotiation* between the social psychologist and the participants as to the structure of an enigmatic episode. Where the social psychologist is a participant observer to an episode, the question of 'what is the *real* Act-action structure?' of an episode can only mean

What description of the Act-action structure can be negotiated most widely between audience and actor (p. 232).

When the social psychologist is not a participant but an observer, empirical checking still takes place in the context of communication: in that case

empirical checking of accounts will be reduced to a negotiation between himself and the participants as to what the Act-action structure is, and perhaps whether there is any Act-action structure present at all. The only possible sense that can be given to the concept of 'the rightness of an account by an ethogenist' is that his account should be the most stable element in a negotiation of accounts, that is the actors will be ready to modify their accounts in the direction of his, when discrepancies exist (p. 232).

In the latter case, negotiation of accounts is conceived of as a kind of mediation:

The context of negotiation can be defined schematically around a minimal three person interaction. There are the two people who are the primary interactants, whose relationship and actions constitute the social phenomenon under investigation. The third person looks on, and negotiates accounts with the primary pair. Each person will be in a position not available to the other two . . . The standard situation of negotiation is typified by a 'family therapy session', such as a man and wife discussing their relationship with the help of a marriage guidance counsellor. The counsellor's job is not just to act as a referee but to enter into the relationship as negotiator of accounts (pp. 236f.).

In the course of this negotiation, the success of the application of models is measured not by correspondence to the results of experimentation, but by the establishing of agreement between the negotiators. It is in the nature of the case that such negotiation may not produce unequivocal results. As Harré and Secord suggest, "there is no possibility of an objective, neutral account by which . . . ambiguities could be finally resolved" (p. 236).[30]

Finally, Harré and Secord follow through the analogy with the realist analysis of the natural sciences to the extent of outlining a counterpart to the metaphysics of powers and natures which Harré had already claimed to be the essential alternative, in the theory of the natural sciences, to the positivist metaphysics of atomistic events. In the realist theory of the natural sciences, the unobservable entities which are components of hypothetical causal mechanisms are conceived as having certain powers to act under given circumstances, and they have those powers in virtue of their natures, which then become the object of further investigation in the next stage of the dialectic of scientific research. In the ethogenic approach, the bearers of powers and natures are taken to be individual human beings. Human powers are analyzed in analogous form to powers of natural entities, namely as tendencies to act in specific ways, under appropriate conditions, in virtue of the specific natures of the individuals. Whereas the action-descriptions contained in powers ascriptions must, following the 'anthropomorphic model of man', be couched in the ordinary language of the individual's social milieu, the description of the intrinsic conditions of the exercise of powers, namely the natures of individuals, are conceived by Harré and Secord as in the form of 'psycho-physical mixes', as basic personality characteristics of individuals, which may, in the future, be analyzable in terms of their physiological basis (pp. 269ff.). In the terms used by Harré in the theory of the natural sciences, this working out of the analogy with the physical sciences seems to betray a prejudice in favour of "a regress of microexplanation", which in the context of social psychology seeks the essential powers and natural kinds which underlie the phenomena of social interaction in the internal constitution of individual human beings, and envisages the eventual discovery of a physiological or biochemical foundation for the powers exercised in interaction.

This brief account of the ethogenic approach to social psychology developed by Harré and Secord will serve as one example of the attempt to transfer the realist theory of science into the analysis of the social sciences. The ethogenic approach is undergoing continual development (Harré, 1977a, 1977b, 1979), and has formed the framework for some empirical research (e.g., Marsh et al.,1978). The present outline, based as it is mainly on the original contribution of Harré and Secord, should however be sufficient to provide a basis for the assessment of this line of argument in the realist theory of the social sciences in Chapter 9.

8.3.2. Realism and Marxism in Sociology

In recent years a number of writers, concerned to develop an antipositivistic yet naturalistic sociology based on the realist conception of science, have found in Marx the model of a realist social science (Keat and Urry, 1975;

Benton, 1977; Bhaskar, 1978b, 1979; Ruben, 1977). Marx's methodology has been interpreted as a more or less adequate fulfilment of the principles of realist science, and his substantive theories as exemplifications of realist canons. An account of the main lines of this interpretation will conclude this chapter on realism in the social sciences.

Keat and Urry describe Marx's methodology as "analogous to that of the realist in the natural sciences" (1975, p. 116), and concede in this judgement that it is necessary to

> avoid giving an interpretation of realism so specific to the natural sciences that no social analysis could possibly resemble it (p. 96).

Like Harré and Secord, therefore, their search for a realist methodology in social science does not amount to a claim for the identity of method between the natural and social sciences, but is rather directed towards working out methods appropriate to the specific subject-matter of the social sciences which are at the same time based on realist general principles. Having rejected the positivist conception of causal laws as being or as based upon regular sequences of observable events, they take as the core of realism the search for causal mechanisms that underlie and generate those observable regularities. Marx's methodology is, they claim, specifically designed to discover such underlying causal mechanisms. In his later writings, his main intention was "to reach a successful characterization of the internal structure of the capitalist mode of production (CMP)".

> More generally he intends to produce descriptions of the structures of various modes: primitive communist, ancient, Asiatic, feudal, capitalist, and so on. He takes the more apparent and observable features of social life to be explicable in terms of these underlying structures. They can be comprehended by the discovery of the causal mechanisms central to each structure; these mechanisms are characterized in terms of the relations between a small number of theoretical entities (p. 96; cf. Benton, 1977, p. 156).

This is an essentially realist and antipositivist intention. However, the structure of a mode of production differs importantly from the objects studied by the natural sciences. According to Keat and Urry, in any mode of production

> there is unity of the elements which comprise each structure. This unity is brought about by a single causal mechanism which is specific to that structure and is its defining characteristic. Second ... the causes and effects of social phenomena are aspects, or applications, of the structural relationships between the elements (p. 97).

In other words, modes of production must be conceived holistically, as complex relationships of elements which only function in the way they do because of their place within the total structure:

> Each element or part is both causally interrelated with every other element and is also a condition of functioning of those elements (p. 102).

The brief outline of historical materialism given in Marx's Preface to *A Contribution to the Critique of Political Economy* (Marx, 1970, pp. 20ff.) is interpreted in this vein. The social relations of production, which form the economic base of a mode of production, are both functionally related to and causally dominant over the other elements of the structure of the mode of production, namely the structures of the superstructure and the relations of exchange, distribution and circulation. At the same time, however, the social relations of production generate internal contradiction between elements of the structure. In the CMP, for example, the social relations of production are essentially those between capitalist and worker. In the structure of the mode of production, both elements, capital and labour, are functionally necessary to the working of the mode, but at the same time the interests of capitalists and workers are structurally antagonistic, and this contradiction results in class struggle. Other contradictions are equally essential, and generate transition from one mode of production to another; the functional relationship between the social relations of production and the forces of production can come into disequilibrium and contradiction through the development of the latter, and when these elements cannot be reconciled within the structure of the existing mode of production, there will be strains, more or less determinative, towards the transition to a different mode of production (Keat and Urry, 1975, pp. 114ff.).

The essential underlying causal mechanism in any mode of production is the nature of the social relations of production and the way they are produced and reproduced. In the case of the CMP, the reproduction of the social relations of production, between the owner of capital and the worker, operates through the specific mechanism which allows the extraction of surplus-value, namely the creation of profit. Marx's discovery concerning the working of this underlying causal mechanism involved the postulation of the theoretical entity, labour-power. Labour-power is a commodity which has the specific characteristic that it is capable of producing more value in the course of a working day than it uses up in the same time. The capitalist buys labour-power at its value, that is its cost of reproduction, and appropriates the product of putting labour-power to work; the difference between the two values is surplus-value, which is the source of the capitalist's profits. The capitalist thereby reproduces his capital, while the worker, paid only the cost of his reproduction, or maintenance, is forced to continue to sell his labour-power; thus the working of the mechanism of the extraction of surplus-value also reproduces the central structural relationship of the CMP, the relationship between capitalist and worker (pp. 106ff.). Similarly, each other mode of production has a central causal mechanism associated with the nature of the extraction of surplus-value.

The other essential difference between the structure of a mode of production and the objects studied by the natural sciences is that the individuals who perform roles in a mode of production also have notions, or theories, of how that mode works and of their role within it, and these notions, or meanings, are also part of the total structure, namely the superstructure. This presents two kinds of problems which are both given realist interpretations. The first of these is that a realist theory of a mode of production must be able to specify the causal mechanisms which lead members of the society in which that mode of production is dominant to have the understanding which they do have of how their society works. Marx thus makes the basic distinction between the appearance, or phenomenal forms, of a set of social relations, and their underlying essence. These need not, and normally will not coincide. The realist interprets the relationship between the phenomenal forms and the essential reality as a causal one: a realist theory must be able to explain the mechanisms by which the structure appears, not as it is, but in the specific way it does, to the members of the society (Benton, 1977, pp. 171ff.). Thus, for the CMP the general nature of the form of appearance of the society is characterized by 'commodity fetishism', to which even the most systematic and disciplined understanding of the working of capitalist society, political economy, is subject (Keat and Urry, 1975, pp. 99f.; cf. Marx, 1976, pp. 163ff.). Since, in the CMP, objects are produced in order to be exchanged, objects of production have not only a use-value but also an exchange-value. These exchange-values 'attach themselves' to commodities. It therefore appears to people living in a society in which the CMP is dominant as if exchange-values were natural and fixed properties of objects, just like the properties of natural, physical things; and the relationships between commodities, which are really the results of social relationships, appear to stand in thing-like relations to each other.[31] The realist interpretation of this appearance must be a causal one; the structure of a mode of production generates causally the way it will appear to the people living under it, rather as the wave length of light causally determines what colour it will appear to those who see it. Of course, not all people will experience a mode of production in the same way, nor will it appear to them in precisely the same way; the realist can therefore suggest that there are causal mechanisms which generate more specific appearances to those who occupy different structural positions within a mode of production. Thus Benton suggests that

> different class ideologies . . . have their material existence in the different practices and experiences of the opposed classes. The source of ideologies is therefore to be sought in the conditions of existence of the classes and their practices themselves (1977, p. 178).

Whereas the capitalist enters the market as a source of finance and as owner of the means of production, the worker, as well as entering the market as a seller of labour-power and as a purchaser of consumption goods, also directly experiences the productive process and the struggles that antagonistic interests in this process give rise to. The relationship between exchange relations and class ideologies is interpreted as the operation of causal generative mechanisms:

> The relationship between real relations and phenomenal forms is, then, a relationship of a causal kind between two aspects of a single structure, one of which has causal (and therefore *explanatory*) primacy over the other (Benton, 1977, p. 178).

For Benton, such a principle is an essential component of a "materialist theory of knowledge", and a natural extension of realism into the social sciences (pp. 171f.).[32]

Secondly, and related to the first problem, the realist can use a scientific theory of the real underlying causal mechanisms at work in a society as a standard against which the notions or beliefs held by members of the society may be judged as more or less adequate. Keat and Urry summarize Marx's theory of ideology in the following terms:

> Marx's professed aim to get to the reality behind the appearances can be seen in two different ways. First, that we explain, in a realist manner, observable features of capitalist

society in terms of unobservable structures and mechanisms. Second, that we criticize members' everyday concepts and understandings because they are based on false theories (1975, p. 180).

Even if it is the case, as Keat and Urry suggest, that "what is problematic is the characterization of both of these in terms of the single distinction between appearance and reality" (p. 180), it is also the case that Keat and Urry see the two issues as essentially connected. For the falsity or distortion contained in the beliefs held by members of a society can, on realist principles, only be judged against a better, more adequate theory, which provides a correct description of the real causal mechanisms at work, instead of false or superficial descriptions of those mechanisms, or, alternatively, descriptions of non-existent mechanisms.

In addition to the causal mechanisms by which the structure of a mode of production generates ideologies which, measured against the standard of a realist theory of those mechanisms, are false or otherwise distorted, the conception of realist structural causality identified in Marx also allows those false beliefs to act as causal conditions for the continued working of a mode of production. Thus the false beliefs generated by the CMP operate in the interests of the dominant capitalist class, functioning to disguise the true nature of exploitation and to legitimate the existing social formation (Keat and Urry, 1975, pp. 178f.). This does not mean, however, that only false beliefs are alike susceptible to causal explanations in terms of the mechanisms against such writers as MacIntyre, that rational and irrational, true and false beliefs alike susceptible to causal explanations in terms of the mechanisms that generate them (1975, pp. 204ff.). Benton too argues that

if scientific knowledge is to be recognised as a reality, distinct from ideological forms of knowledge, then we need a theory of the mechanism of its production . . . the theory of the mechanism by which scientific knowledge is produced must involve some theory of the transformation of ideology into science (1977, p. 173; cf. Bhaskar, 1978b, p. 20).

Benton considers and rejects Althusser's use of Bachelard's concept of 'epistemological break' as a way of conceptualizing the mechanism of the production of scientific knowledge, suggests that detailed research into the generation of scientific knowledge in the case of different individual sciences would be necessary to make progress on this problem, and hints that the mechanisms of institutionalization of scientific research would have to be studied through the concepts of the relations of production and the relations of distribution of knowledge. Consistently, Benton also argues that

> in establishing a conceptual break with the prevailing ideological formations a scientific theory is necessarily subversive, and is a threat to the ideological domination of the prevailing ruling class (1977, p. 191).

Whereas false or distorted beliefs contribute to the maintenance of a mode of production, true or scientific beliefs may, as in the case of the natural sciences from the seventeenth century onwards, also be required by the ruling class, who will therefore have to balance their need for science with the dangers posed by its subversive potential. Thus explanations of the development of true knowledge take, for the realist, the same form as explanations of the generation of false or distorted appearances, namely, the discovery of the causal mechanisms which generate such beliefs.

One of the arguments by which Keat and Urry reject the view that only false or distorted beliefs can be given causal explanation involves another component of their interpretation of realism. They argue that among the causes of people holding 'rational', 'true', or 'undistorted' beliefs may be considered the reasons for which they hold them (1975, pp. 208ff.). This derives from the more general position, that among the causes of actions which, on a realist interpretation, may be components of the mechanisms which generate actions, the reasons which people have for acting must figure prominently. As we saw above, Keat and Urry rejected the antinaturalist argument that reasons for actions could not be considered to be causes, as based on a positivistic conception of causality. Now we can see that this negative critique is followed by the positive thesis that agents' reasons for action are central components in the causal mechanisms by which actions are generated. For the realist, the causal link between reason and action is a crucial link in the mechanisms by which a structure of social relationships is reproduced, since the position which a person occupies in such a structure provides them with beliefs and theories which generate reasons for acting in ways which reproduce the structure itself (Keat and Urry, 1975, pp. 192ff.; cf. Bhaskar, 1978b, p. 12).[33] For structures of social relationships, although according to the realist 'real' and possessed of an ontological status in their own right, can only continue to exist by being reproduced in the action of individual people. Since people act in accordance with the meanings that the situations in which they find themselves have for them, these meanings and the reasons for action which they support must for the realist form part of the causal mechanisms by which social structures are reproduced.

This does not mean that people *intend* to reproduce social structures through their action. Bhaskar, for example, sums up 'the relationship between society and people' by suggesting that

> people, in their conscious human activity, for the most part unconsciously reproduce ... the structures that govern their substantive activities of production. Thus people do not marry to reproduce the nuclear family, or work to reproduce the capitalist economy. But it is nevertheless the unintended consequence (and inexorable result) of, as it is also the necessary condition for, their activity (1978b, p. 16).

People also 'occasionally transform' the structures in which they live, but this too need not be consciously intended. Presumably the transition, for example, from the 'feudal' to the capitalist mode of production was nobody's conscious intention. But Keat and Urry recognize that, in the Marxist tradition, the transition to a socialist society is supposed to be an exception to this general statement; in this context they also make a few remarks on the realist conception of socialist society. In a socialist society

> the social world will not be seen to consist of things or to be thing-like in constitution; it will be clearly the result of human activity; and people will correctly feel that collectively they both have changed and can change it. But social relations will still have ontological status and causal power, although people will not ascribe erroneous status to such relations ... There will clearly be social phenomena external to each individual and partly determinative of individual action. Such 'social facts' will include the social organization of production, language, norms of appropriate socialist behaviour, and so on (Keat and Urry, 1975, pp. 194f.).

Since social relations and structures of social relations will still have their own 'reality', it will still be possible and indeed necessary to develop social scientific theories on realist lines of the mechanisms by which such structures operate, not so much because people will otherwise have generally false or distorted conceptions of socialist structures (for although some people will have false or distorted views, these will neither have their source in the structural features of the society nor be necessary conditions for the continued existence of the structure, as in the CMP), but rather for 'technical' purposes: since a socialist society will have to be "highly organized, industrialized, and rationalized", realist theories of its structure will be needed "to indicate exactly how its members can intervene in its working so as to control and transform its constitutive relationships" (1975, p. 195).

Finally we turn to the problem of the empirical testing of models of the causal mechanisms through which social structures are generated and reproduced. As Harré and Secord did in the case of ethogenic social psychology, realist interpretations of Marxist social science must also face the question of how such models of underlying causal mechanisms can be justified or validated. Each of the writers we are discussing turn their attention to this question. Keat and Urry raise it as the question of how models of modes of production may be applied to actual societies, but their answer is somewhat vague. They suggest that, in order to "identify the degree to which any mode of production dominates a given society", it is necessary to have "criteria to detect the empirical presence of the given theoretical entities to be found in Marx's analysis" and to know "how to examine the relationship between two or more modes present in a society" (1975, p. 116). This does present the testing of a model of the structure of a mode of production in a given social formation as analogous to the testing of existential hypotheses in the natural sciences, but little further information is given on how this analogy may be implemented. Keat and Urry also realize that, if the falsity of ideological conceptions of a social formation is to be measured against the truth of a realist model, criteria for the superiority of the realist model have to be available. But their response to this problem is the rather weak statement that, in comparison with the situation in the natural sciences, the task of getting agreement in the social sciences when faced with competing theory-dependent descriptions of a social structure "is more problematic" (1975, p. 180).[34]

Bhaskar is only a little more informative on this question. Following his own analysis of the natural sciences, he argues that the ability of natural scientists to check the models they propose depends on their ability to isolate, in the laboratory, individual causal mechanisms as 'closed systems'. Social scientists, on the other hand, are unable to do this: closed systems do not occur spontaneously in societies, nor can social scientists artificially create them. The implication drawn from this is that

> the social sciences are denied, in principle, decisive test situations for their theories. This means that the criteria for the rational confirmation and rejection of theories in

social science *cannot be predictive*, and so must be *exclusively explanatory* (1978b, p. 19; cf. 1979, p. 58).

However, apart from the suggestion that the special position of the social sciences requires them to generate 'real definitions' of forms of life that have already been identified and described through the categories of the members of the society whose forms of life are being studied (1978b, pp. 21f.), no further discussion is given of what these 'explanatory' criteria might be, or how they might be used to adjudicate between competing theories of social structures.

Benton addresses himself in more detail to the problem of the validity of theoretical descriptions of underlying causal mechanisms. He does not accept that there can be any timeless criteria for the validity of scientific knowledge, and rejects the Kantian concept of the 'thing in itself', or Althusser's version of it, the distinction between the 'real object' and the 'object of knowledge', as idealist, and hence incompatible with realism or materialism (1977, p. 186). Instead he poses the question of validity in more historical terms, as the problem of comparing at any given time two theories claiming to explain the same subject-matter. If all description is theory-dependent, the problem arises of knowing whether competing theories even refer to the same objects. Using the example of Lavoisier's chemical revolution, Benton argues that continuity of subject-matter despite change in theoretical language is established through 'identity of reference' of theoretical concepts; that is, both Lavoisier and Priestley could recognize the concepts 'dephlogisticated air' and 'oxygen' as referring to the same substance, and this identity of reference

is dependent upon the existence of procedures for producing the substance which may be taught, learned, copied, etc. without presupposing what is at issue between the two chemical theories (1977, p. 187).

Given that identity of reference is thus established in this (apparently) operationalist manner, one theory will be superior to another if the later theory can cope with the anomalies thrown up by the earlier theory. Each theory will have internal criteria of validity, which specify what kinds of causal mechanism may be considered 'plausible', and which underlie the tests of existential hypotheses concerning theoretical entities and their interrelationships. But when it turns out that these internal criteria of validity cannot cope with areas of phenomena which are resistant to explanation in their terms, strain is generated which is resolved by transition to a new theory, whose concepts have identity of reference with those of the old theory, but whose internal criteria of validity are able to cope with the areas anomalous in the old theory (1977, pp. 194ff.).[35] Internal criteria of validity, precisely because they are internal and theory-dependent, cannot be used to adjudicate between different theoretical systems. Nor can they serve as standards for the development of new theoretical systems. Benton argues that the only criteria which can be used to judge the validity of the knowledge produced within a theoretical system as a whole are the overarching criteria of logical non-contradiction and consistency.

The materialist thesis of the existence of a real world, prior to and independent of knowledge of it, together with the realist conception of science as knowledge of the nature, structure and constituent causal mechanisms of that reality, imply a conception of the unity of scientific knowledge such that the ontologies and basic concepts of each science and branch of science will, to the extent that they express genuine knowledge, be mutually consistent (1977, p. 197).

Since, as Benton argues, sciences develop at different rates, the more developed sciences provide for the less developed a set of more or less fixed points of reference with which they must strive to bring themselves into accord. In this way the "normative conception of the unity of the sciences" exercises a "critical function in the development of scientific knowledge" (p. 198).

In none of this discussion on the question of validity does Benton make any concession to specific features of the social sciences. One must assume that he holds that theoretical concepts in the social sciences refer to real objects in just the same way as those in the natural sciences, and that 'identity of reference' of the concepts of successive theories is established in just the same way too. Indeed, he argues that

only through some notion of the unity of the sciences such as the one I have advocated (which is one fundamentally different from that of the positivists) which includes the social sciences, can there be any rational foundation for the notions of progress and objectivity in the social sciences. This in turn justifies the practice of looking to the natural sciences for analogues of the conceptions of causality and of explanation which are required in the social sciences, and also of demanding of social-science theories that they are consistent with the basic laws and propositions of the physical sciences (p. 199).

It appears that the basic criterion for the choice between competing theories in the social sciences is, according to Benton, that they should be consistent with the findings of the natural sciences.

It is clear from the discussion in this section that there is considerable divergence between the implications drawn from the realist theory of science by sociologists with Marxist orientations and those drawn by proponents of the ethogenic approach to social psychology discussed earlier. These concern the strength of the analogy between conceptions of 'reality' in the social and the natural sciences, and the nature of empirical testing in these two branches of knowledge. There is also a certain degree of diversity even among the small group of Marxist-oriented sociologists whose work has been the subject of this section. Clearly there is considerable uncertainty as to the precise implications for the social sciences of the critique of positivist naturalism from the standpoint of the realist theory of science. This will need to be borne in mind in the final chapter of this part, in which the three different forms of antipositivism in the theory of the social sciences will be brought into relationship with each other.

CHAPTER 9

DISCUSSION: ANTIPOSITIVISM IN THE PHILOSOPHY OF THE SOCIAL SCIENCES

In the first three chapters of this part, approaches to the theory of the social sciences have been presented from the three standpoints which understand themselves to be opposed to 'positivism', critical rationalism, critical theory, and scientific realism. As in the corresponding chapter of the previous part, Chapter 5, the task of the present chapter is to bring these three approaches into relation to each other, in order to assess their relative claims to advance adequate antipositivist principles for the social sciences. In the course of the argument it will be possible to draw on the preliminary conclusions reached in the earlier chapter on the relationships between these three approaches in the context of the philosophy of the natural sciences.

The argument of this chapter will be restricted mainly to a discussion of the relationship between critical theory and scientific realism. The reason for this has already been mentioned in Chapter 1. The so-called *Positivismusstreit* in German sociology, as it developed from the meeting in Tübingen in 1961, was constituted as a confrontation between critical rationalism in the shape of Popper and Albert and critical theory in the shape of Adorno and Habermas. As such it has been the object of considerable analysis at various levels of detached reasonableness and vigorous polemics. Similarly, this time predominantly in the British literature, there has been a certain amount of discussion of the respective merits of 'Popperianism' and realism for the development of the social sciences. Apart from one or two exceptions, however, the argumentative contact between critical theory and scientific realism has scarcely existed, a product of the relative mutual insulation of German critical theory and Anglo-American sociology which is still only beginning to dissolve. The main aim of this work is to clarify some of the confusion surrounding the variety of attacks on 'positivism', and in particular to bring into relation with each other the antipositivisms of critical theory and scientific realism. Hence the focus of this chapter will be on the 'third side of the triangle'.

The other two sides of the triangle cannot, however, be ignored, and some aspects in particular of the conclusions which might be drawn from the *Positivismusstreit* have already been mentioned and may at this point be summarized. Critical rationalism, although from the point of view of critical theory incapable of producing an adequately reflective theory of knowledge, does go part way towards such reflection by treating methodological rules, not just as either neutral descriptions of what scientists do or as intuitions of the essence of science, but as conventions, whose adoption is open to critical discussion. In the context of the natural sciences, this methodological conventionalism appeared, to critical theorists at least, to be in practice inconsequential, since the range of potential sets of methodological rules available for adoption is limited by the demands that the results of the natural sciences must be in such a form that they should be transformable, in principle if not in practice, into technically applicable rules of instrumental action. In the context of the social sciences, on the other hand, alternative sets of methodological rules appear to be given by the existence of the tradition of the hermeneutic *Geisteswissenschaften*, which offer the social sciences an alternative model of procedure, based on the understandability (*Verständlichkeit*) of symbolically structured systems of action and cultural institutions. Critical rationalism thus seems to be faced with material which, on its own terms, would allow a reasoned discussion as to which, if either, of these sets of methodological rules would be more appropriate to meet the goals set for the social sciences, a discussion which would avoid the dogmatic anti-philosophical prejudices of (logical) positivism. However, critical rationalism cuts short this potential discussion with the aid of two assumptions, on the basis of which it embraces the doctrine of the unity of scientific method: firstly, that it is the aim of any science to describe the relationships which actually obtain in reality; secondly, that in the social sciences, just as in the natural sciences, these relationships take the form of invariant regularities which are to be captured in propositions expressing universal laws. From the point of view of critical theory, the first of these assumptions contains a doctrine of metaphysical realism, which in the social sciences just as much as in the natural sciences fails to do justice to the problem of the constitution of areas of reality through synthetic activity, and which also fails to raise the question of the meaning of the ascription of independent reality to differently constituted object-realms. The second assumption is also held by critical theory to be unjustified, on the grounds that in the realm of culture, which is historically variable and open to transformation by human activity, no universal laws expressing invariant regularities can be found. We have seen that, while on critical rationalist principles it can never be known for certain that any hypothesized law actually is universal, the existence of numerous and well attested candidates for laws of nature in the natural sciences as they have developed over the past centuries makes it implausible to doubt that methodological rules advocating the search for laws of nature can indeed be implemented; this conclusion holds even if the analysis of what laws of nature are has to be amended on realist lines. In the social sciences, on the other hand, there are no generally accepted candidates for the status of universal law. While critical rationalists might be correct to argue that critical theorists cannot know for certain that there are no universal laws in the sphere of social action and social institutions to be found, the lack of any generally accepted candidate for such a law adds strength to the fall-back position available to critical theory, that the criterion of hermeneutic understandability can be used to judge whether any social regularity can be fruitfully investigated, not from the standpoint of its possible universality, but from the standpoint of its potential transformability and accessibility to criticism.

Faced with this critique, critical rationalism has had to pay more attention to the hermeneutic alternative. However, it has refused to put into question the two assumptions stated in the previous paragraph, from the standpoint of which it rejects hermeneutics as pre-scientific, indeed anti-scientific. Thus Albert writes of the current of hermeneutic thought that in it

> there culminates a process of resolute subjectivization of thought, which involves a turning away from the posing of epistemological questions and from the methodical style of the natural sciences, and in which the relevance of rational argumentation and objective orientation for cognition is greatly diminished (1971, p. 51; cf. 1975).

As we have seen, Albert understands hermeneutics exclusively in the historicist or quasi-theological form in which it is also rejected by critical theory, and however apt the charge of 'anti-modernism' and of undifferentiated anti-scientific prejudice might be when applied to some aspects of the Heideggerian tradition, this cannot in general be said of the use made of hermeneutics by critical theory. If the objections of critical theory to critical rationalism's metaphysical realist and nomological assumptions are justified, critical rationalism is also deprived of its defence against another principle of hermeneutics. For it is only by assuming the existence of universal laws in the social and cultural realm that critical rationalism can hope for a 'neutralization' of the values inherent in the concepts used for social description. Popper has long argued that in science the choice of concepts is unimportant: concepts are more or less useful for the formulation of nomological hypotheses, and those which turn out not to be fruitful can be rejected in favour of others (e.g. 1972, pp. 123f.). Values intruding into the object-language may thus be neutralized by the discovery of laws with some degree of explanatory power. If it cannot be assumed that such laws can be found, this neutralizing force is lost, and the hermeneutic principle reappears that the choice of a conceptual scheme for the description of social and cultural relationships always remains bound to initial 'pre-scientific' pre-judgements, which can never be dissolved by objectivist stratagems, but which can at best be more or less consciously articulated, and more or less adequately justified.

This assessment of the balance of argument in the *Positivismusstreit* may be borne in mind as we turn to the relationships between critical theory and scientific realism. The major conclusion drawn in Chapter 5 concerning the evaluation of scientific realism from the standpoint of critical theory was that scientific realism must be judged to be a new form of 'objectivism'. However great an insight into the methods of the more advanced natural sciences might have been achieved by scientific realism, it is incapable as it stands of incorporating this insight into a comprehensive reflective theory of knowledge. It is unable to reflect the constitution of an object-realm in which the methodology it describes may be meaningfully applied, but believes that this methodology is essential to any scientific grasp of 'reality', relying on a doctrine of metaphysical realism which is not fundamentally different from that of critical rationalism, however different their conceptions of scientific method might otherwise be.

As has already been hinted, the implications of this objectivism for a realist theory of the social sciences are far-reaching. An objectivist theory of science, which believes that the scientific knowledge it analyzes provides knowledge, however theoretically mediated, of reality as such, has no conceptual tools available by which it might study the conditions in which methodologies incorporating different categorial frameworks could appropriately be applied. Having rejected the principles of deductivism as an inadequate, as well as philosophically confusing, description of the methods of the more advanced natural sciences, realism has no way in which it is able to justify the transference of this critique of 'positivism' over to the social sciences. It can, quite successfully, point out that a considerable amount of both antinaturalist and naturalist argumentation in the philosophy of the social sciences is based on a mistaken conception of the natural sciences, and this form of argument is particularly successful against more blatant examples of empiricist naturalism, whose proponents find themselves in the position of the members of a community who treat as a rule-book to govern their own activity not just an ethnographic account of another community, but a mistaken ethnographic account of another community. But it is unable to give convincing reasons either for or against the adoption of a modified naturalism, which has learned the lessons of the critique of deductivism.

The absence of a reflective theory of knowledge provides a clue to the variety of forms taken by realist theories of social science which has already been noted. As we have seen, in some cases realists argue that the adoption of an antipositivist, realist naturalism does not mean that the methods of the natural sciences must be slavishly followed in the social sciences. Even within the natural sciences there are considerable variations of method, and, as Bhaskar puts it, in the social sciences too

> it is the nature of the object that determines the form of its science (1978b), p. 2; cf. Benton, 1977, p. 137).

Without any theory of the constitution of object-realms by synthetic activity, the delineation of differences in the objects of different sciences must proceed in an *ad hoc* manner. Thus the specific differences which Keat and Urry, for example, find in the subject-matter of the social sciences, namely the structural interrelationship of elements in holistic social structures, are quite different from those which Harré and Secord stress, namely the generation of social action by the conscious self-monitoring activity of social agents. In each case, the method for studying this specific subject-matter is, it is claimed, derived from a realist conception of science, in that it is designed to get 'behind' patterns of observable events to the 'generative mechanisms' beneath; but in each case the representation of the kind of mechanisms which could be sought with the use of iconic models is based on an analogical extension of the idea of a causal mechanism to spheres of phenomena whose specific features are loosely recognized but not adequately grounded in a theory of knowledge. The senses in which the central structural mechanism of a mode of production or the rule-role structure of an episode are 'like' (or 'unlike') the causal mechanisms studied by natural sciences cannot be theoretically analyzed by scientific realism without abandoning the objectivist doctrine of metaphysical realism in which it is caught.

Critical theory, in contrast, bases its opposition to any doctrine of the unity of scientific method, as we have seen, on the argument that two quite different object-domains are constituted in the spheres of instrumental and communicative action. These object-domains are accessible to two fundamentally different kinds of experience, sensory experience or perception, and communicative experience or understanding. It would follow from the thesis of a lack of a reflective theory of knowledge in scientific realism that realism would be unable to appreciate the force of this distinction, any more than any other objectivist, and in that sense positivist, theory of science. This appears to be the case.

The inability of scientific realism to assimilate the distinction between sensory and communicative experience makes itself manifest in a variety of ways. The clearest of these concerns the problem of the status of unobservable, theoretical entities in the social sciences. In the natural sciences, as we have seen, an important part of the realist argument rests on the idea that causal mechanisms differ in degree of accessibility. In the case of accessible mechanisms, hypotheses concerning them can be directly tested by inspection. In the case of quasi-accessible and inaccessible mechanisms, on the other hand, it is necessary for scientists to use their imagination to construct hypothetical models, based on analogy with other better known mechanisms, of mechanisms which might be responsible for the observable events under study. Such models will postulate the existence of entities, inaccessible to direct observation, whose ways of acting produce observable phenomena. Such entities may be called theoretical, and one distinctive feature of scientific realism, in contrast to descriptivism and instrumentalism, is the interpretation of theoretical terms in the natural sciences as hypothetical references to unobservable, but nonetheless existing entities, whose existence may be indirectly checked by the use of scientific instruments to detect their effects. It can immediately be seen, in the light of previous discussion, that 'unobservability' in this context takes its meaning by reference to the capacities for sensory experience present in the human organism. 'Observable' mechanisms and entities are those which are accessible to the human senses, either unaided or modestly extended. 'Unobservable' mechanisms and entities are those which are inaccessible to the human senses, even when modestly extended, but whose effects can be made accessible with the use of scientific instruments, which extend the range of sensory experience into areas beyond the reach of the human senses. The categorization of theoretical entities as unobservable is thus bound to the sphere of sensory experience; the transfer of the language of observability and unobservability to the sphere of society and culture, access to which according to the hermeneutic principles accepted by critical theory is by means of communicative experience, is therefore fraught with difficulties. For structures of social relationships are not 'unobservable' in the same sense as elementary particles or black holes; it is not that they are inaccessible to human sensory experience, requiring instead sociological equivalents of bubble chambers and radio telescopes in order to be perceived, but rather that they are only accessible to a form of experience which goes beyond sensory experience, namely communicative experience.

Keat and Urry, however, despite their brief outline of Habermas' distinction between instrumental action and communicative interaction and the correspondingly different cognitive interests in each sphere (1975, pp. 222ff.), do not appear to notice any discrepancy between the theory of knowledge which these distinctions contain and their use throughout the rest of their argument of the terms 'unobservable' and 'theoretical entity' to refer to social structure (1975, pp. 94f., 96f., 124f., 159f., etc). They argue that structures are theoretical entities in that they are not perceivable by direct observation. The notion of structure is explicated with reference to the structure of a building:

> When we refer to the structure of a building we are not referring to an observable property such as height or size. A structure is not something that can be directly perceived by our senses. The concept of structure involves abstraction from the observable properties of any such building (1975, p. 120).

When the concept of structure is transferred to the social sciences, it is possible to distinguish positivist and realist interpretations:

> For the positivist to talk about the social structure is to talk of the laws and regularities found to obtain between observable phenomena. For the realist a structure consists

of the systems of relationships which underlie and account for the sets of observable social relations and patterns of social consciousness (1975, p. 121).

Keat and Urry here collapse two distinctions into one. In the first case they distinguish between 'direct' sense perception and 'abstraction' from sense perception. Even to entertain such a distinction seems to be a relapse into a strict empiricist model of perception as pure receptivity, and one might say that all sense perception involves some degree of 'abstraction' in the sense that representations must be synthesized under categories. However, they do not seem to realize that, despite this 'abstraction', the perception of the structure of a physical object such as a building still takes place through the medium of sensory experience. Access to social relations and structures of social relations, on the other hand, takes place through the medium of communicative experience. But Keat and Urry fail to see the significance of this second distinction, and believe that the notion of 'abstraction' from 'observable' social relations is sufficient to explain how information about social structures is acquired.

As we have seen, one basic element of the distinction between sensory experience and communicative experience is, according to Habermas, that the former is 'monologic' while the latter is 'dialogic'. These terms characterize the relationship of the knowing subject to the object of research, and the form of the pragmatic context in which that relationship is embedded. Realist writers, again because their lack of a reflective theory of knowledge prevents consideration of the conditions in which object-domains are constituted, are unable to see that the communicative nature of hermeneutic understanding constitutes an essential difference between the natural and the social sciences and a fundamental block to the naturalism they unreflectively embrace. Keat and Urry's discussion of this problem is again typical. They distinguish between 'explanatory understanding' and 'interpretive understanding' (1975, pp. 167ff.). Explanatory understanding they represent as the aim of all sciences, being the kind of understanding one has of something when one knows the explanation for it. Interpretive understanding, on the other hand, is the type which some hold to be specific to the social sciences, and involves understanding the linguistic or more generally symbolic meanings which are present in agents' actions and beliefs. This form of understanding presupposes the existence of a shared language or symbolic system between the researcher and the agents whose actions, beliefs and social relations are being studied. However, Keat and Urry argue, the necessity of this form of understanding does not constitute an argument for antinaturalism. For the analysis of the natural sciences also requires reference to interpretive understanding. All knowledge-seeking enterprises are social in nature, and take place in scientific communities.

One fundamental feature of such communities is the presence of linguistic communication between its members. Without this, scientific activity would be impossible. But what is involved in this communicative interaction is the interpretive understanding of each scientist by others. Thus, to account for the possibility of explanatory understanding, we must provide a philosophical analysis of the nature and status of the interpretive processes, the shared linguistic and contextual meanings, upon which this is based (1975, p. 175).

Keat and Urry do recognize an 'asymmetry' between this interpretive understanding within a scientific community and the interpretive understanding which social scientists aim at, but do not see in this asymmetry any grounds for antinaturalism. It is clear, however, that they have failed to grasp the significance of the distinction between the dialogic nature of the relationship between subject and object in the social sciences and the monologic nature of this relationship in the natural sciences. It is of course true that a scientific community in both cases presupposes the existence of shared media of communication between its members, but only in the social and cultural sciences can these media of communication also be shared by the 'object' of study. Thus only in the social and cultural sciences do the problems of hermeneutic interpretation arise, which according to the critical theorist make impossible the use of 'pure' monologic languages for the description of social and cultural affairs.

Precisely the same failure to appreciate the real basis of critical theory's antinaturalism appears in the recent writings of Anthony Giddens, despite his much greater familiarity and sympathy with the work of Habermas. Giddens represents the specific features of the social sciences, in contrast with the natural sciences, by means of his concept of the 'double hermeneutic':

The theory-laden character of observation-statements in natural sciences entails that the meaning of scientific concepts is tied-in to the meaning of other terms in a theoretical network; moving between theories or paradigms involves hermeneutic tasks. The social sciences, however, imply not only this single level of hermeneutic problems, involved in the theoretical metalanguage, but a 'double hermeneutic', because social-scientific theories concern a 'pre-interpreted' world of lay meanings. There is a two-way connection between the language of social science and ordinary language (1977, p. 12; cf. 1976, p. 162).

Giddens uses this concept of the double hermeneutic to criticize Habermas' analysis of the natural sciences and hence of the relationship between the natural and social sciences. For if "hermeneutic problems are as basic to science as to more 'sedimented' traditions", then the natural sciences too are concerned with 'understanding':

Science is certainly as much about 'interpretation' as 'nomological explanation' ... 'Explanation' in science is most appropriately characterized as the clarification of queries, rather than deduction from causal laws, which is only one subtype of explanatory procedure. In scientific analysis 'why-questions' are normally answered by rendering a phenomenon intelligible or meaningful within the context of a paradigm or theory. Recognition of the hermeneutic character of scientific theories and their mediations shows that science is oriented in a fundamental way to 'understanding' (1977, p. 149).

But if this is the case, then Habermas' distinction between the empirical-analytic sciences based upon a technical interest and the historical-hermeneutic sciences based upon a practical interest is unsatisfactory:

Hermeneutic problems cannot be at the same time confined to one class of disciplines, and yet also span them all. If those fields of study (as Habermas agrees) which concern human action involve what I call a 'double hermeneutic', the hermeneutic mediation of meaning-frames (paradigms) must be regarded as posing central problems for any epistemology of natural science that seeks to go beyond the discredited formulae offered by logical empiricism (1977, p. 151).

Habermas' 'positivist' conception of the natural sciences thus leads him to an oversimplified antinaturalism that merely reproduces "much of the old *verstehen/erklären* opposition" (1977, p. 150).

That Habermas' conception of the natural sciences retained too many elements of 'positivism' was, of course, one of the main conclusions of Part II of the present work. However, Giddens, like Keat and Urry, carries this re-interpretation of the natural sciences too far. Although he is quite aware of the distinction between monologic and dialogic experience (1977, p. 139), he dissolves the distinction by insisting that the natural sciences too are concerned with understanding. He thus disguises the fact that the relationship between members of natural-scientific communities and their objects of study is indeed necessarily monologic; there can be no dialogue, or communicative interaction, between people and physical objects. People can impose meanings on physical objects, but these cannot be shared by the objects themselves. Objects cannot participate in the relationship which people have with them, they are rather the potential objects of monologic instrumental manipulation. By concentrating on the hermeneutic aspects of communication within scientific communities, Giddens takes the focus off this monologic manipulation, and ignores the fact that science proceeds by experimentation, which is itself a form of technical action. The relationship of social scientists with their object of study, however, is a more systematic form of ordinary language communication, and to the extent that social scientists must employ hermeneutic procedures, they are engaged in dialogue whose regulative idea is mutual understanding. The use of the term understanding in the context of the natural sciences is therefore misleading, for although in a general sense it is perfectly meaningful to talk about understanding how physical systems work, the specific sense in which the social scientist understands the symbolically structured products of human action requires a dialogue relationship. This is made clear, as Apel points out (1977, p. 431), in the use of the term '*Verständigung*', which encompasses both the understanding of meaning and the process of achieving agreement; the term cannot be used in relation to physical objects, precisely because they are not potential partners to a dialogue.

A similar mistake occurs in Keat's recent and more thorough analysis of Habermas' theory of knowledge (1981). Keat attempts to show that Habermas' theory of cognitive interests cannot perform the task demanded of it, of showing that different methods of inquiry are appropriate to differently constituted object-realms. He argues that Habermas can give no reason why hermeneutic methods are appropriate only to an object-domain of meaningful entities, constituted by the practical interest, and not to the object-domain of natural objects. For

> if these two object-domains differ solely by virtue of their different constitutive interests (operating, presumably, upon the same uncategorized, homogeneous, 'externality'), what grounds can there be for rejecting a 'hermeneutics of nature', for *not* 'choosing' to constitute it via the practical interest? (Keat, 1981, p. 79.)

If, alternatively, these object-domains are 'ontologically' distinct, Habermas must abandon the doctrine of constitutive interests.

It is quite clear that Keat has misunderstood part of Habermas' argument. For Habermas does not envisage the technical and practical interests constituting different object-domains out of an undifferentiated reality. Quite to the contrary, he shows that the object-domain of the cultural sciences confronts the investigator as *already* constituted in the sphere of communicative action by the actors' categories of ordinary language (Habermas, 1972, pp. 192f.). The researcher gains access to this object-domain by means of communicative experience, by dialogue. But this is not the case with the object-domain of the natural sciences, to which access is gained by monologic, sensory experience. These quite different conditions of knowledge are what Habermas appeals to in arguing for the appropriateness of different methodological frameworks in the natural and the cultural sciences. Keat, like other realist naturalists,[36] misses the force of this distinction, and thus builds up his criticism of Habermas' epistemology on a false premise.

Among the realist writers whose work has been discussed, the only ones to show any real insight into the importance of the distinction between the monologic nature of sensory experience and the dialogic nature of communicative experience are Harré and Secord. We have already seen how they locate the process of empirically checking hypothetical models of the structure of episodes in a relationship of negotiation between social psychologist and the participants to the episode, and they stress at a number of points that the process of research in the social sciences entails the researcher entering into a social relationship with the 'objects' of study, and that "the social psychological investigation is itself a social episode" (1972, p. 238). They also recognize the specifically hermeneutic problems this communicative relationship raises for traditional naturalist conceptions of science; the interpretations which social psychologists produce are rivals with the already present interpretations of the participants, and there is no completely neutral ground on which the scientist could stand that would avoid the need for what Harré and Secord term 'negotiation'. They also remark on the absence of hermeneutic conditions in the natural sciences: 'obviously one cannot negotiate with atoms' (1972, p. 236). Yet even they finally attempt to heal the breach they have created between the study of nature and of society in their determination to show that common features of scientific research apply equally to both. For they assimilate the negotiability and re-negotiability of interpretations of social episodes to the 'essential revisability' of truth claims in both the natural and the social sciences, concluding that 'the demand for final, absolute unrevisable truth cannot be met in either field' (1972, p. 236). Using the terminology developed by Habermas in more recent works, it can be seen that Harré and Secord here confuse the rules for the argumentative testing of truth claims in discourses, which Habermas stresses must be the same in different sciences (1973a, p. 172), with the constitution of object-domains of physical objects and symbolically pre-structured meaning-systems, accessible to sensory and communicative experience respectively. It is the latter distinction which Harré and Secord had originally noticed; but since their commitment to metaphysical realism forbids them a theory of knowledge which could reflect the conditions in which object-domains are constituted, their insight is lost and falls victim to their insistence on the principles of the realist theory of science.

We can see, therefore, that realism's objectivism leads to its inability to justify its own version of naturalism, to a misapplication of the realist theory of the status of theoretical terms, and to a failure to appreciate the specific nature of communicative experience and of hermeneutic interpretation and understanding, which constitutes a central feature of critical theory's opposition to naturalism and of its conception of the foundation of both the historical-hermeneutic and the critical social sciences. One particular consequence of the objectivism of scientific realism in contrast to the reflective theory of knowledge of critical theory can be seen in their respective interpretations of Marx. Critical theorists, as we have seen, find in Marx's method of ideology-criticism one of the basic models of a critical social science, but they also identify in Marx a 'latent positivism' which attempts to conceive the theory of historical materialism in general, and the theory of the development and necessary transformation of the capitalist mode of production in particular, as scientific theories with the same epistemological status as the natural sciences. Scientific realists, on the other hand, interpret Marx's historical materialism as a realist social science, which correctly attempts, more or less successfully, to transfer the principles of realist science to the study of social structures, their development and transformation, while avoiding the pitfalls of positivist naturalism. However, it is not at all apparent that realists, by interpreting Marxism as 'realist science' rather than as 'positive science', thereby escape from the criticism which critical theorists have directed against the latter.

Central to this criticism is the problem of how the application of the conceptual framework of Marxism is to be justified. As we have seen in the discussion of critical rationalism, the conceptual framework of a natural science is justified by its fruitfulness in generating hypotheses, and stabilized by the success of these hypotheses in standing up to experimental test; concepts embodied in hypotheses which withstand testing are retained in the conceptual framework of a science, while those which do not are rejected. This state of affairs must be the same for science interpreted on realist principles. Concepts referring to types of unobservable, theoretical entities, for which it is claimed that they are not merely useful assumptions but that they actually exist, will only be retained in the conceptual framework of a science if the entities to which they refer can be shown to exist; a paradigm case of this stabilization of the conceptual framework of a science on realist lines would be, perhaps, the abandonment of the concept of 'phlogiston' and its associated concepts and the stabilization of the concept of 'oxygen' and the associated concepts of Lavoisierian chemistry. That this problem of the stabilization of the conceptual frameworks of sciences is central to the realist theory of science is shown by the systematic place accorded to the categories of 'real essences' and 'natural kinds' in the more Aristotelian-oriented versions of the realist theory which were stressed in the previous part of this work. For the systematic purpose of these categories is to demonstrate that the conceptual frameworks of the natural sciences are not the arbitrarily chosen affairs that the nominalist presuppositions of 'positivism' would have them be, but rather, by a process of the formulation of real definitions, approximate ever more closely to the structure of reality itself, describing the real essences of the natural kinds of things of which reality is formed.

It would follow from the hermeneutic presuppositions of critical theory that this procedure for the stabilization and justification of conceptual frameworks cannot be the same in the social and cultural sciences. Social scientists are also social individuals, and are embedded in the systems of social action which they study. The conceptual frameworks they use themselves have social significance to participants in the systems of communicative action which are the subject-matter of the social sciences. The conceptual frameworks used in the interpretation of systems of action are tied to the initial position of the interpreters and their hermeneutic pre-judgments, and in the absence of either well-attested invariant laws on the one hand or generally accepted real definitions of natural kinds on the other, the results of research carried out on the basis of these pre-judgments cannot be unambiguously fed back into the choice of conceptual frameworks to guarantee their stabilization. As Habermas puts it, "we are practically interested in society", and therefore "a prior understanding originating in interested experience always infiltrates the fundamental concepts of the theoretical system" (1974, p. 210). It is open to Marxism, as to any other theoretical system in the social sciences, to elucidate and attempt to justify the prior understanding that underlies its conceptual framework, but it cannot short-circuit this procedure with the argument that only its concepts reflect the structure of reality.[37]

Interpreting Marxism as critical theory, this prior understanding is a philosophy of history based on a materialist reconstruction of the Hegelian theme of alienation and its *Aufhebung* or transcendence (Wellmer, 1971, pp. 85f.). The potential exists in the human species to make history consciously according to its collectively formulated will. But the process of all history up to the present has been a quasi-natural process,[38] made up of the activity of individual human beings but under conditions over which they have had no control and therefore with consequences which none of them intended. Although individual people are active subjects, the human species is not the conscious and active subject of history as such (Horkheimer, 1972, p. 200; Habermas, 1974, p. 247). The reason for this, Marx argued, lay in the alienation of labour (Marx and Engels, 1968, p. 45). This domination of dead labour over living labour, while manifesting itself as the constraint of objective, quasi-natural forces, is not a natural, inevitable necessity; in fact, in the capitalist mode of production, the highest form of alienated labour, it produces the conditions of its own transcendence, among them unprecedented wealth and the impoverished industrial proletariat, which is capable of and has an interest in the abolition of alienation and the creation of a new form of society. This materialist philosophy of history underlies and forms the starting point for the later more detailed conceptualization of the workings of the capitalist mode of production; the theory of surplus value, for example, remains bound to this hermeneutic pre-judgment of history as the history of the alienation of labour (see Ollman, 1971, pp. 176ff.).

The foundation and justification for the conceptual framework of Marxism is thus bound to this philosophical pre-understanding, which itself takes its justification from the position of the proletariat. Habermas interprets Marx as follows:

> Alienated labour, the domination of dead labour over the living, as expressed in the pauperized existence of this class, developed into a commanding need for its own abolition as a class – a need that was 'the practical expression of necessity'. Therefore, so Marx argues, at the same time as this class has occupied the objective position of the proletariat within the process of production it has also achieved a point of view outside this process, from which the system as a whole can be comprehended critically and convicted of its obsolescence (1974, p. 249).

Philosophically, the proletariat was accorded the status of the universal class, the class whose interests were universal and not particular as those of all previous classes had been; the implementation of the interests of the proletariat would therefore require the transcendence of alienation, the abolition of 'natural' society, in which there is a cleavage between the particular and the universal interest. Epistemologically, the social position of the proletariat gave them the vantage-point from which capitalist society could be viewed critically with respect to its transformability. Practically, the proletariat was the only social force objectively capable of carrying through the transformation of capitalism and bringing about communism. In this way the historical location of the proletariat provides epistemological justification for the conceptual framework used by Marx to analyze capitalist society. At the same time, as we have seen, that analysis was addressed to the proletariat; it was supposed to contribute to the critical self-reflection of the proletariat, to provide it with the categories through which it could understand its historical location, criticize the ideologies which disguised and legitimated capitalist societies, and develop theories and practical strategies of revolution.

The interpretation of Marxism as a realist science takes little if any more account of this problem of providing epistemological justification for the conceptual framework of Marxism than does the 'positivist' Marxism which is one of the main objects of criticism for critical theorists. Bhaskar adopts the essentially critical rationalist criterion:

> like any other fundamental conceptual blueprint or paradigm in science, historical materialism can only be justified by its fruitfulness in generating projects encapsulating research programmes capable of generating sequences of theories, progressively richer in explanatory power (Bhaskar, 1979, p. 53).

Keat and Urry briefly mention Lukács' formulation of the problem, grounding the truth of knowledge gained from the standpoint of the proletariat in the universality of its interests, but appear to dismiss it and to pass on to an analysis of the concept of reification (1975, pp. 180ff.). But without Lukács' aid, which is so cursorily rebuffed, they have, as we have seen, no satisfactory solution to offer. Benton, following Althusser, rejects the attempt to ground the conceptual framework of Marxism in the theory of alienation as 'humanism', from which Marx is supposed to have broken decisively in his later writings (1977, pp. 146ff.). Since Benton does not recognize the possibility of a form of theory which is, in Habermas' sense, 'between' science and philosophy, and thus does not consider critical theory as a distinct type of knowledge, the only alternative to the interpretation of Marxism as 'philosophical humanism' is Marxism as science. The justification for the conceptual framework of Marxism must therefore be identical in form to the justification of the conceptual framework, or as Benton terms it 'theoretical system', of any science. No general epistemological justification can be given, Benton claims, but in a specific context a theoretical system can be superior to another on the grounds of its consistency with other, more advanced sciences, and on the grounds that there exist "procedures for identifying, recognizing and producing the entities, processes and structures to which the basic concepts of a theory refer" (1977, p. 198). The first of these criteria is so loose that its application to the social and cultural sciences would seem to have no differentiating power. Since Benton rejects, with good reason, positivist versions of the unity of science which would make the social sciences logically deducible from the physical sciences, the criterion of logical consistency can be satisfied by any social scientific theoretical system which does not actually contradict established natural laws. The second criterion contains the realist assumption that the theoretical concepts of the social sciences refer to 'theoretical entities' in the same way as those of the natural sciences. But the reflexivity of ordinary language, to which the concepts of the social sciences are indissolubly bound, precludes the formulation of unambiguous demonstrative and recognitive criteria for the referents of theoretical concepts, whose selection and significance remain hermeneutically linked to "an anticipatory interpretation of society as a whole" (Habermas, 1974, p. 210). This difficulty is simply not recognized by Benton, and it is fatal to his attempt to ground the concepts of historical materialism on realist principles. Realism, while eschewing as humanism an overall conception or philosophy of history which could provide an explicit hermeneutic starting-point for the more specific conceptual framework of Marxism, is denied any alternative approach to the problem by its own doctrine of metaphysical realism.

Realism's objectivist conception of Marxism as science, with the same cognitive status as the natural sciences, generates in addition precisely those interpretations of historical materialism which critical theory characterizes as 'latent positivism'. This is most apparent in realism's approach to the theory of ideology, which will be discussed in more detail in the final part of the present work. It is already manifested, however, in the realist treatment of the base-superstructure metaphor. The general recognition of the inadequacy of economic determinism is shared by realists, and Keat and Urry follow an interpretation which attempts to reconcile the dominance of the economic, "determination in the last instance", with a conception of the functional incorporation of the base and the superstructure into a total system of mutually interrelated elements. Once this has been done, however, it no longer makes any difference if it is claimed, as it is by Keat and Urry, that

> social classes . . . are not merely elements of the economic base; class relationships are importantly determined by the superstructure (1975, pp. 108f.).

For the interpretation of the relationship between the base and the superstructure as one of functional interdependence in a dynamically self-regulating system already makes the realist vulnerable to criticism from the critical theorist. As Wellmer argues, such a conception must treat the 'contradiction' between the productive forces and the relations of production as a systemic disequilibrium to which there is a technically specifiable solution in terms of a reorganization of social relationships more suited to the continued development of the productive forces (1971, p. 93), which accords closely with Keat and Urry's realist interpretation of structural contradiction (1975, pp. 101, 114ff.; cf. Benton, 1977, pp. 156ff.). And it must treat class conflict as the necessary consequence of such a systemic disequilibrium, according it the functional role of impelling the disequilibrium towards a resolution (Wellmer, 1971, pp. 94ff.), which again is consistent with realist accounts of class conflict (Keat and Urry, 1975, pp. 101, 107ff., 115f., 136f.; Benton, 1977, p. 160). The realist search for causal mechanisms, together with the systemic conception of base-superstructure relationships, leads realism to interpret class conflict as the direct outcome of systematic contradiction; the critical theorist, on the other hand, while not denying that class conflicts on an ideology-critical approach too "would be those of *material life*, because they would be grounded in the discrepancy between the possible fulfilment and the actual suppression of real needs", insists that they must also be conceived

as the protest of repressed individuals against forms of domination whose legitimations had become questionable, and against the burdens and deprivations demanded by such forms of domination and experienced as capricious and unjust (Wellmer, 1971, p. 96).

Class conflict, in other words, would be seen not as merely causally generated by structural antagonism, but also as the outcome of a 'critical dissolution of false consciousness', which is located not at the level of systems of instrumental action but in the sphere of the reflective criticism of ideology spurred on by the experience, including the material experience, of domination. The inability to separate out an ideology-critical science, operating in the sphere of the self-reflective criticism of quasi-natural systems of domination, from a science directed toward causal explanation, even when this is interpreted on realist rather than positivist principles, thus mars realism's interpretation of Marxism, and leads it to reproduce the elements of 'latent positivism' which critical theorists had located in Marx's own thought.

The objectivism of the doctrine of metaphysical realism thus leaves scientific realism, as a theory of the social sciences, vulnerable to the full weight of the antipositivist critique of critical theory. We may finally turn to a discussion of the implications of the use of the term 'real' in the theory of the social sciences, which will parallel the analysis of the use of this term in the theory of the natural sciences advanced in Chapter 5. That there is a problem associated with the application of the term 'real' to the subject-matter of the social sciences is recognized by scientific realists themselves. Keat and Urry discuss the problem in the context of the concept of 'reification'. Marx, in his remarks on commodity fetishism, had pointed out that, in a society whose economic relationships were based on the production of commodities, social relationships between people take on "the fantastic form of a relationship between things". From this, Lukács elaborated the concept of reification, in which a relation between people takes on the character of a thing and thus acquires a 'phantom objectivity', an autonomy that seems so strictly rational and all-embracing as to conceal every trace of its fundamental nature: the relation between people (1971, p. 83).

Reification was a central feature of capitalist society and a symptom of human alienation, but it was not inevitable, and it was the task of the critique of political economy and of Marxist analysis in general to unmask reification in theory and contribute to its dissolution in practice. Reification, from this point of view, was a sin which any social theory which envisaged the transformation of capitalist society must strive to avoid. Yet, as Keat and Urry suggest, the realist proposal to treat social relationships and social structures as 'real' might appear to open realist social science, and especially realist Marxism, to the charge that it necessarily commits that sin. Keat and Urry are therefore concerned to provide an analysis of the nature of reification, and to specify realism in such a way as to avoid those elements of reification that should indeed be avoided.

To this end, and drawing on the work of Berger and Pullberg (1966), they distinguish five elements of reification:

The first is the claim that there is nothing that is real which does not have the character of a *thing*; given then that social phenomena are real, they are taken to be things ... Second, the products of human activity are seen as having an *independent existence* separate from the activity that produces them ... Third, reification involves the ascription of *ontological* status to social roles and institutions, whereas we should only ascribe such status to persons ... Fourth, such roles and institutions are taken to determine *causally* how human beings behave; society is seen as producing men and not vice versa. ... Fifth, reification involves the claim that humans are *powerless* to change their social and historical circumstances (Keat and Urry, 1975, p. 185).

Keat and Urry then sketch out the relationship of realism to each of these five elements. It is accepted that it is a mistake to take social phenomena to be things, in the sense of material or physical objects, but denied that realism is committed to that mistake; for not only does realism accord no privileged status to ordinary 'macro-sized observable objects', it is also the case that

realism, as a methodological position relating to the nature of scientific theories and explanations, involves no *a priori* commitment to a physicalist ontology. Realism, in the sense we have given to it, is independent of metaphysical disputes between, for example, materialism and dualism (1975, pp. 191f.).

Secondly, Keat and Urry argue that while some products of human activity do have independent existence separate from the activity that produces them, namely material objects, this is indeed not the case with other categories of human products, such as social institutions, roles, and symbols (1975, p. 186). Yet although it is true that social phenomena do not have an existence separate from the totality of human activity, "for any single person such relations have an existence apart from his or her activity" (1975, p. 193). This will continue to be the case even in a socialist society which has otherwise transcended elements of reification. It is thus not a mistake for realism to recognize this. Thirdly, it is not mistaken reification to ascribe ontological status to social relations and social institutions, but these do not have the same ontological status as either material objects or individual persons.[39]

Fourthly, on a realist account social relations do have causal influence on the course of human behaviour, but do not completely determine it, since the subjective meanings which people ascribe to situations and actions also contribute to the determination of human behaviour. Moreover, the nature of this determination varies between different types of society (1975, pp. 193ff.). Here Keat and Urry appear to be suggesting that the category of reification is too undifferentiated for the analysis of the causation of action, and that therefore realism need not feel burdened with this element of the charge. Fifthly, they deny that the first three elements of reification imply the fifth, since human beings are not even powerless in the face of natural forces. They may nonetheless feel powerless if independent reality is ascribed to social relationships, and to the extent that single individuals cannot change the structure of social relationships, this feeling of powerlessness is justified. Collectively, however, human beings can transform the social structures they live in, and in a socialist society they will know that they have changed society and are capable of changing it again. None of this is challenged by a realist theory of society (1975, pp. 185, 193f.).

However, other realist writers diverge from Keat and Urry's conclusions. Benton, for example, does not allow a sharp separation of realism from materialism, but rather interprets realism as part of a "materialist theory of knowledge" (1977, p. 171). A materialist theory of knowledge must "recognise the reality of the object of knowledge, independent of the 'knowing subject', the process of production of knowledge, and the knowledge itself"; among such realities are 'thought', 'ideas', and 'knowledge' itself. Benton therefore proposes that 'ideologies' have a 'material reality' of their own, a "real existence, in the sense of having some contributory effects on the maintenance or overthrow of a social order" (p. 162, cf. pp. 177ff.). This appears to contradict Keat and Urry's statement that "social relations are not material things but they are real" (1975, p. 193). This may, however, be due to a divergence of usage of the term 'material', for while Keat and Urry use it in the sense of a physicalist ontology, Benton repudiates such a usage as metaphysical (p. 171; cf. Keat and Urry, 1975, p. 192). That Benton has qualms about the use of the term 'material' to refer to ideologies is suggested, nevertheless, by his occasional use of scarequotes (1977, p. 175). What is clear, however, is that Benton is far less concerned about the problem of reification than are Keat and Urry. He defends Durkheim's conception of the facticity of the social world against the charge of reification, and rejects the 'humanist' philosophy which insists on an essential difference between reified and non-reified social forms (1977, p. 93). He does not raise the question of the relationship between the 'real existence' of social structures and the human activity which produces or sustains them. And he would presumably dismiss Keat and Urry's distinction between radically different 'ontological statuses', such as may be ascribed to material objects, persons, and social institutions, since he insists on the 'mutual consistency' of the ontologies of all sciences (1977, p. 197). It would not be going too far to say that Benton seems not to have noticed that the use of the term 'real' in the social sciences might pose any difficulties.

Bhaskar, on the other hand, does attempt to analyze what is involved in the ascription of 'reality' in the context of the social sciences. In the study of nature, he argues,

> science employs two criteria for the ascription of reality to a posited object; a perceptual and a causal one. The latter turns on the capacity of the entity whose existence is in doubt to bring about changes in material things. It should be noticed that a magnetic or gravitational field satisfies this criterion, but not a criterion of perceivability (1978b, p. 4; cf. 1979, pp. 15f.).

In the case of social structures and social relationships, the first of these criteria is inapplicable, since "society . . . is necessarily unperceivable" (1978b, p. 18). But the second, causal criterion can, according to Bhaskar, be applied to establish the reality of social structures:

> if it can be shown that but for society, certain physical actions would not be performed, then . . . we are justified in asserting that it is real (1978b, p. 16; cf. 1979, pp. 49ff.).

Bhaskar claims that Durkheim, in invoking the criterion of constraint as part of his definition of social facts, was in effect using this causal criterion to establish the reality of social facts. He quotes Durkheim as follows:

I am not obliged to speak French with my fellow-countrymen nor to use the legal currency, but I cannot possibly do otherwise. If I tried to escape this necessity, my attempt would fail miserably. As an industrialist, I am free to apply the technical methods of former centuries; but by doing so, I should invite certain ruin. Even when I free myself from these rules and violate them successfully, I am always compelled to struggle with them. When finally overcome, they make their constraining power sufficiently felt by the resistance they offer (Durkheim, 1964b, p. 3).

Bhaskar comments:

Durkheim is saying in effect that but for the range of social facts particular sequences of sounds, movements of bodies, etc. would not occur (1979, p. 50).

Thus, according to Bhaskar, to ascribe 'reality' to social structures, social relationships, and other irreducibly social phenomena is to point to their causal effects on human behaviour, and these causal effects make their presence manifest through the experience of *resistance* which they offer to human intentions. At the same time as suggesting that this causal criterion for reality can be used equally in the social as in the natural sciences, Bhaskar also stresses fundamental differences in the ontological status of social and natural structures, which he sees as posing essential limitations on the possibility of even a realist naturalism:

(i) social structures, unlike natural structures, do not exist independently of the activity they govern;

(ii) social structures, unlike natural structures, do not exist independently of the agents' conceptions of what they are doing in their activity;

(iii) social structures, unlike natural structures, may be only relatively enduring (so that the tendencies they ground may not be universal in the sense of space-time invariant) (1978b, p. 14; 1979, pp. 48f.).

These features are clearly connected to those which Keat and Urry discuss in relation to reification, but whereas they do not see them as posing problems to the project of a realist naturalism, Bhaskar offers them as limitations to that project, limitations which prevent the methods of the natural sciences from being incorporated without significant modification into the social sciences.[40]

The objectivism of scientific realism stamps all these contributions to the discussion of the nature of the 'reality' of social phenomena, a question which must lie at the heart of any realist theory. This objectivism, in the shape of a doctrine of metaphysical realism, attempts to conceptualize an independent reality and develop a theory of its fundamental features without raising the question, central to a critical philosophy, of the constitution of that 'reality' through synthetic activity. The structuring of reality by fundamental categories must be traced back to capacities of the knowing subject, which, if they are not the capacities of any consciousness whatsoever, as Kant conceived them to be, must, if they are to have any objectivity, be derived from the fundamental conditions of existence of the human species. As we have seen, Habermas, by reflecting on the "fundamental conditions of the possible reproduction and self-constitution of the human species" (1972, p. 196), discerned three sets of such conditions which generate orientations from which reality can be constituted: the need to exist in a natural environment and the capacity for increasingly reliable instrumental manipulation of that environment; the need for symbolic interaction between members of the species and the capacity to combine individuation of experience with the maintenance and improvement of mutual understanding and consensus in communication; the capacity for rational reflection on the conditions of development of the individual and the species, which makes possible the dissolution of rigidities which have become entrenched in earlier phases of that development. To these three sets of conditions, which Habermas conceptualizes in terms of three fundamental interests which guide the learning process necessary to the continued development of the human species, correspond the basic categories according to which reality can be constituted and known. The lack of a constitution-theory of reality allows scientific realists only to speculate concerning the nature of potentially different spheres of reality which might be accessible to scientific investigation, but not to provide epistemological foundations for the categorial schemes under which this 'reality' may be studied.

Yet, as we have seen in Chapter 5, the critical theorist need not simply reject realism entirely, but can reconstruct it within a constitution theory. In the case of the realist theory of the natural sciences, Schnädelbach re-worked the remark of Popper, that the reality of nature is experienced through 'resistance', into the theory that the ascription of independent reality to natural objects and structures takes place under conditions of reflection on the experience of obstruction or resistance to human instrumental action. Following Peirce, he argued that 'reality' should be understood as the object of true belief, that which, in the limit, a community of investigators would stabilize their belief around, earlier beliefs having proved unstable because action based on them proved unsuccessful. But the category of action, whose lack of success forces reflection on the conditions of failure and inquiry into the beliefs on which unsuccessful action was based, must itself be specified; and in the case of the natural sciences, whose method of inquiry itself involves instrumental action, this category is the category of technical or instrumental action.

We must now ask whether the same reconstruction of realism can be implemented in the case of the other categories of methodological inquiry, the historical-hermeneutic and the critical sciences, distinguished by Habermas' epistemological theory. First we may notice the similarity to Popper's conception of the experience of the resistance offered by reality which is presented by Bhaskar's conception of the resistance of social reality, in which he draws on Durkheim's criterion of constraint. Yet if we examine the examples of constraint or resistance given by Durkheim, it is clear that the nature of the resistance varies between the different cases, and all of Durkheim's examples differ from the resistance which Popper has in mind. Popper, it may be recalled, gave as an elucidation of the resistance offered by reality the example of "the experience of a blind man who touches, or runs into, an obstacle, and so becomes aware of its existence" (1972, p. 360). It is by generalizing from this case of physical obstruction that Schnädelbach could argue that this was a constitution of reality from the standpoint of the resistance it offered to instrumental, or technical, action. Yet when Durkheim suggests that, although he is not obliged to speak French with his fellow-countrymen, he "cannot possibly do otherwise", he is not alluding to a physical resistance, nor is the action which is constrained by this resistance categorizable as instrumental action. It is rather, and most clearly in this example, communicative action. Durkheim presumably means that if he does not use the medium of French, or some other shared symbolic system, in interaction with his partners in dialogue, communication cannot take place. As Schnädelbach expresses it,

an experience of resistance in the context of communicative action is therefore the experience that particular elements of action may be problematized only at the cost of a destruction of the continuity of communication (1972, p. 109).

One 'runs into' reality, in the sphere of communicative action, when one realizes that communication is being endangered or disturbed, in contrast with the possibility of an uncoerced mutual understanding and agreement (*Verständigung*). This latter idea acts as the criterion of success in communicative action, just as reliable technical control acts as the criterion of success in instrumental action. The experience of failure in each case triggers off reflection on the conditions of this failure and of possible success, which in the case of communicative breakdown demands inquiry into the conditions of possible mutual understanding, and generates the categorial framework of the hermeneutic sciences, just as failure in instrumental action generates the categorial framework of the natural sciences.

This pragmatistic interpretation of the resistance presented to Durkheim's communicative action by the requirement that he speak French may be contrasted with a different interpretation which parallels the doctrine of metaphysical realism. A realist could argue that communication obstacles can only be overcome if the participants to a dialogue used words in their correct meaning: this would parallel the realist's argument in the theory of the natural sciences, that success in technical manipulation is dependent on a correct understanding of the real connections in nature. Schnädelbach suggests that exactly this thesis has been argued by 'conceptual realism', which proposes that there exist unchangeable or ideally 'true' meanings of linguistic expressions, divergence from which prevents true communication (1972, p. 107). Only when words are used in their correct meanings can what is said correspond to what is meant. But this thesis cannot be acceptable. Instead the idea of a successful communication must be seen as the regulative idea for the attempt to narrow dissensus over the meanings of linguistic expressions, guided by the ideal goal of a final consensus.

As in epistemological realism, which appeals to the 'independence' of action-elements involved in instrumental action in order to arrive at a definition of the expression 'empirical truth', the thesis of realism in the realist doctrine of ideal meanings rests on an ontologization of that which would be the content of a final consensus (1972, p. 108).

In the case of communicative action, however, the restrictions on the possibility of arriving at consensus are constituted not, as in the case of instrumental action, by the structure of the physical world, its causal laws or real essences, but by the inescapable conditions which must be met for dialogue to be possible.[41]

Not all obstacles to consensus in the sphere of communicative action are to be identified with the 'reality' which, in the limit, would constrain the consensus of any community of investigators. For such obstacles also take the form of symbolically structured social and cultural formations, which, since they are the product of human historical activity, are in principle, or so it may be assumed, further transformable by human activity. This is true in the case of the other examples of constraint or resistance advanced by Durkheim in his definition of social facts. Two kinds of social fact are mentioned by Durkheim, in addition to what we have interpreted as the need for a shared medium of communication. The first of these are 'purely moral maxims', 'the conventions of society'; the second the conditions of economic action. In the first case, constraint is exercised by the sanctions which operate in society to ensure conformity with normative rules, though of course Durkheim's analysis of sanctions and of constraint shifted from one in terms of the external application of positive and negative sanctions to one in which the system of normative rules is internalized by the individual in the course of socialization, so that the constraint operates, as it were, from within (Durkheim, 1964b, pp. 2f.; cf. 1961, and Parsons, 1949, pp. 378ff.). In the second case, constraint operates in a quasi-natural manner, by the operation of the 'laws' of the market. Thus when Durkheim points out that

> As an industrialist, I am free to apply the technical methods of former centuries; but by doing so, I should invite certain ruin (1964b, p. 3),

it must be understood that the ruin will come about through the failure of Durkheim as industrialist to observe the conditions of competition. Now Durkheim himself does not make any sharp distinctions between these two cases, nor between them and the first case discussed, the necessity for a shared medium of communication, although he does, in later writings, make a much clearer distinction between these social forms of constraint and the operation of natural constraint, distinguishing between the moral authority of normative rules and the utilitarian considerations involved in relation to the control of our adaptation to natural forces, such as a disease (1961, pp. 29ff.). But such distinctions must be made, for the nature of the resistance offered by such social facts differ between themselves, and not just from the resistance of nature.

Thus neither the normative rules of a society, nor the conditions of economic action, offer resistance of such a kind that they constitute inevitable limits to the possibility of consensus-formation. Both are the historical product of human activity, and both can in principle be transformed. The experience of resistance which they may engender is thus not the experience of a 'reality' which must either be brought under technical control or adapted to, but the experience of conditions which are open to transformation. Nor is the experience of resistance merely a disinterested experience of an external facticity; it may rather be the experience of conditions of material existence which have become intolerable, or of a normative system which has become questionable. In other words, it may be an experience of a resistance which can give rise to criticism, a criticism which is rational precisely because the conditions which present resistance are not inevitable conditions which constrain any possible consensus, but transformable conditions which stand in the way of consensus. It is therefore an experience of resistance which can generate the basic categories of critical social sciences.

The form of resistance which constitutes the central categories of critical social theory is that of domination (Habermas, 1972, p. 313).[42] Domination can operate through the medium of overt coercion and force, but in advanced industrial societies it is more often stabilized by legitimation. Legitimation may derive either from the normative power of cultural traditions or from the quasi-natural power of the conditions of economic action; in the first case domination is made to appear right and just, in the second it is made to appear technically inevitable. Critical social theory, guided by an interest in practical emancipation, which is provoked by the experience of repression or domination, is directed at the critique of such legitimations, and aims to show that cultural traditions which shore up domination and inequality cannot be based on 'true' consensus,[43] and that apparently inevitable conditions of economic action are only quasi-natural objective powers, not to be mistaken for natural forces. The final consensus towards which critical social theory is in the limit oriented is an uncoerced consensus, in which the members of society, freed from the constraints of domination, will be able to decide collectively the form their society should take under the given material conditions which are available to them.

When Bhaskar, following the principles of the realist theory of science, suggests that the experience of resistance is the criterion of the reality of society, he is thus taking only a partial step away from the objectivism of realism, just as Popper only took a partial step away from the objectivism of positivism.[44] The 'reality' of nature is constituted by reflection on the ultimate constraints to human instrumental action presented by the enduring structure of natural things. The 'reality' of social relationships, on the other hand, is constituted by reflection on the the resistance presented to the formation of a just and so far as is possible equal society by structures of domination brought into existence in the quasi-natural course of previous human history. Unless it can be shown, as it has not yet been shown, that there exist necessarily enduring structures in society as there are in nature, this latter 'reality' can not be assumed to incorporate untranscendable limits to the achievement of consensus, apart from those conditions which must be met for communicative action of any kind to be possible. Only a realism thus transformed and reconstituted can meet the demands of critical theory's antipositivism.[45]

It is not clear that realism has any convincing reply to such a critique. In this respect the situation is different in the theory of the social sciences from that in the theory of the natural sciences. In the latter case it could be argued that, despite its objectivism, realism had identified and incorporated into a theory of scientific method features of research in the natural sciences which deductivist approaches to the philosophy of natural science had failed to comprehend. These features, central among them the invention of hypothetical models of inaccessible generative mechanisms and the checking of such models by the testing of existential hypotheses, themselves derived, as it was argued in Chapter 5, from striking advances in the development of the material forces of scientific inquiry. Critical theory, in developing a reflective theory of knowledge which would provide rational foundation for the methods of the differing branches of systematic inquiry, had, by accepting the deductivist principles of explanation in the natural sciences, itself failed to take account of the advances which realism in its new forms made the basis of its theory of science. To that extent the formulations of the realist theory had at least to be taken note of, and reformulated into a more comprehensive and differentiated theory of the natural sciences. But the development of the material forces of inquiry are necessarily bound to the sphere of sensory experience. They are scientific instruments which serve to extend sensory experience beyond the range of the human organic sensory equipment. In the sphere of communicative experience, in which the object of the social and cultural sciences are constituted, there appears to be no direct analogue to the development of the material forces of inquiry, on which an analogous realist theory of social scientific method could be founded.[46] The only suggestion which realists have put forward as an analogy with scientific instruments, namely Harré and Secord's suggestion that the dramaturgical standpoint is to social psychology what microscopes are to physical science, itself rests on the human capacity for self-monitoring action, and there seems to be no apparent sense in which this capacity could be said to undergo development comparable with that of the material forces of inquiry. Realism in the social sciences is therefore based either upon a set of very weak analogies, which have no epistemological significance, or upon a dogmatic insistence on the unity of scientific method, comparable in its lack of reflection to the similar insistence of critical rationalism.

This conclusion also implies that the proposal tentatively advanced in Chapter 5 for a framework within which historical development of the method of the natural sciences could be conceptualized, cannot be directly transferred to the sphere of the social sciences. The history of the social sciences cannot be understood in terms of the development of the material forces of inquiry used in them, for there is no such development. Nor can the history of the social sciences be related directly to changes in the form of the division of labour, since the social sciences are not concerned exclusively with the forms of organization and institutionalization of instrumental action, but also with the forms of organization of patterns of communicative action outside the sphere of work, however widely the latter is conceived. To inquire into alternative general conceptual frameworks for a more historical understanding of the methods of the social sciences would take us too far afield for the present work. We must rest on the essentially negative conclusion, that in the light of criticism from the standpoint of critical theory, scientific realism has little to offer to the understanding of the social sciences, and should confine its endeavours to the source of its original insight, the sphere of the natural sciences.

PART IV

POSITIVISM, ANTIPOSITIVISM AND IDEOLOGY

CHAPTER 10

POSITIVISM, ANTIPOSITIVISM AND IDEOLOGY

During the West German *Bundestag* election campaign of 1976, the Christian Democratic Union issued an advertisement, one of a series, consisting of a photograph of a boffin-type scientist considering the question: "*Warum CDU wählen*" ("why vote CDU"), and below the photograph his answer:

> . . . *weil wir unsere Zukunft nur mit einer ideologiefreien Wissenschaft bewältigen* (Der Spiegel, 13.9.1976, p. 61). (. . . because we can only handle our future with science free from ideology.)

This curious attempt to fight an election with the weapons of epistemology followed considerable debate in the West German press concerning what might be called the fourth *Methodenstreit*: the setting up by a group of professors of a study group, financed by the Thyssen Foundation, with the name '*Wissenschaftforschung in der Bundesrepublik*' ('Science research in the Federal Republic'), in order to combat what they perceived as political challenges to the autonomy of science (Hübner *et al.*, 1976). Among the challenges they identified and discussed at a meeting held in the Munich Hilton in March 1976, was that stemming from the work of the project group 'Alternatives in Science' at the *Max-Planck-Institut zur Erforschung der Lebensbedingungen der wissenschaftlich-technischen Welt* at Starnberg, under the direction of Jürgen Habermas and Carl Friedrich von Weizsäcker. This group had developed the concept of 'Finalisierung' (finalization), with which they argued that once sciences had reached a certain level of maturity, they generated alternative research programmes which made possible decisions as to their desirability, not on purely internal criteria of scientific development, but on criteria of social value; science policy could thus be opened up to processes of democratic decision-making (Böhme *et al.*, 1972, 1976). This challenge to the autonomy of science was interpreted by its critics, among them leading critical rationalists, as the intrusion of ideology into science, a formulation which the CDU sharply picked up in its advertisement (see Pfetsch, 1979).

These incidents serve to remind us that the issues discussed in the previous two parts under the rubric of the critique of positivism are not purely epistemological, but also sociological and political. It has already been pointed out that critical theory's critique of instrumental reason operated simultaneously on an epistemological and a sociological or social-theoretical level, although in the foregoing discussion the emphasis has been on the former. The concept of ideology is central to critical theory's sociological critique of positivism, just as it is also central to critical rationalism's defence of the freedom of science. For whereas critical rationalism sees the importation of heteronomous interests into science as the degeneration of science into ideology, critical theory sees the technical or instrumental rationality, which is an essential element of 'positivist' theories of science, as the core ideology of advanced industrial societies. In the course of the *Positivismusstreit*, both critical rationalists and critical theorists designated the theories of the others as 'ideological' (Adorno *et al.*, 1976, pp. 30, 63f., 256), even if by this term they meant very different things. Thus the attempt, outlined in the first part of this work, to create conditions for mutual understanding between these traditions by translating them into comparable epistemological terms, runs up against the barrier of mutual accusations of ideological distortion.

However, a detailed consideration of all the implications of this overlapping of epistemological and sociological problems would overstep the bounds set for the present work, not least those of space. Instead, this concluding part will examine the concepts of 'ideology' and its relationship to 'science' advanced by each of the three schools of thought which form the object of this study. As Kreckel has pointed out (1975, p. 92), the fact that the concept of ideology straddles both epistemology and sociology accords it a key position in the understanding of the debate between critical rationalism and critical theory, and a further assessment of the theories of ideology developed by scientific realists will enable us to locate realism within the range of problems thrown up by the *Positivismusstreit*.

All genuine theories of ideology [47] contain a concept of ideology which

belongs to the set of theoretical attempts to uncover the general causes of error, of false consciousness; as far as reflection about knowledge can be followed back in the history of philosophy, there can always be found answers to the question of what dispositions or mechanisms hinder the knowledge of the truth (Schnädelbach, 1969, pp. 73f.).

Thus all theories of ideology contain as a contrast concept a concept of truth, and a conception according to which the disclosure of ideology is a process of enlightenment. What distinguishes theories of ideology is what meaning is given to the notion of truth and what significance is accorded the process of enlightenment. Each of the three schools of thought under study, critical rationalism, critical theory and scientific realism, have developed theories of ideology which can be analyzed in these terms.

CHAPTER 11

THE CONCEPT OF IDEOLOGY IN CRITICAL RATIONALISM

According to the correspondence theory of truth advocated by critical rationalism, truth can only be discovered by systematic scientific research. By a process of methodical testing of falsifiable hypotheses, which embody conjectures about what nomological relationships exist in reality, scientific research is able to approximate its theories, in a quasi-asymptotic way, to the truth. Truth is never finally attained, but rather approached by theories with a higher degree of explanatory power than their predecessors, a higher truth-content and a lower falsity-content (Popper, 1972, pp. 46ff.).

In order that this gradual approximation to the truth should be possible, a number of conditions must be satisfied, which relate to the falsifiability of proposed hypotheses. Falsifiability is not an intrinsic feature of a theory; rather any theory can be rendered unfalsifiable by the adoption of 'immunization strategies', or what Popper earlier referred to as 'conventionalist stratagems', which protect a theory from potential falsification by, for example, redefining the concepts which appear in it, reformulating the background assumptions which have to be made in order that a theory may be applied to a concrete situation, or revising the theories of measurement in which sensory observations are theoretically conceptualized (Popper, 1968, pp. 81ff.; Albert, 1968a, p. 30). Critical rationalists therefore adopt, as a methodological convention which should facilitate the progress of science towards the truth, the rule that theories should not be protected from falsification by immunization strategies. When this rule is not in force, so that the conditions for the falsifiability of hypotheses do not hold, the possibility that error will not be exposed is always present. Critical rationalists can therefore use the term 'ideology' to cover any thought which is protected, by any of a range of possible immunization strategies, from critical test (Albert, 1968a, pp. 86ff.).

From this concept of ideology there follows a specific conception of enlightenment, or ideology-criticism. According to Albert, the ideology-critical task of critical rationalism is

> to diminish the irrationality of social life by means of the fruitful application of the results and methods of critical thinking to the formation of social consciousness and thereby also of public opinion (1968a, p. 88).

In this task

> it is often less a question of the transmission of detailed knowledge, than of the teaching of methods which would enable the individual to develop an independent judgment and thereby to see through strategies of immunization, of obscuration, of screening, of mystification, in short, dogmatic modes of procedure (1968a, p. 88).

Enlightenment can only come about by the increasingly extended application of the methods of critical thinking, as these are understood by critical rationalism. The theory of ideology is thus essentially an epistemological theory, which treats all forms of thought as subject to the same logical principles, and which can uncover ideological thought in any department of human belief, whether it concerns the understanding of natural processes or social ones.

> 'Enlightenment' on this approach is therefore synonymous with the ideology-critical condemnation of all modes of thought which do not measure up to the model of the natural sciences (Kreckel, 1975, p. 94).

It would appear as if there were no particular sociological content to this 'scientistic' conception of ideology and its criticism, and that the theory could not be used for sociological analysis.

However, critical rationalists can argue that whether the conditions for the application of the methods of critical testing are present itself depends on social factors. Whether the conditions for the exposure of belief to potential falsification are satisfied depends on the way in which science is institutionalized in society. For it is necessary that scientists should be free to subject to critical scrutiny any mode of thought which might be performing an immunizing function and preventing a belief from possible rejection. This recognition of the social conditions necessary for critical reason to operate is expressed in Popper's conceptions of the 'open society' and of the 'public character of scientific method', by means of which he describes the necessity for the institutionalization of freedom of criticism in the scientific community (Popper, 1962, ii, pp. 217ff.). These conditions are not automatically given, however, and critical rationalists have learned from the Marxist tradition that some beliefs in particular have the function of legitimating prevailing social relationships characterized by inequality and domination: such beliefs, which serve to stabilize unequal and unfree societies and prevent the discovery of conditions in which inequality could be reduced and freedom increased, will be, it is argued, especially likely to be protected from public criticism and shielded by immunization strategies. Ideology-criticism on critical rationalist principles will have to be directed particularly against such beliefs, for their very social function generates the suspicion that truth is being hidden by such immunization (Münch, 1973, pp. 151ff.). This specific application of ideology-criticism nevertheless remains bound to critical rationalism's original epistemological concept of ideology, for it involves the critical testing of the cognitive status of inequality-legitimating beliefs, following the same methodological principles that would be applied to any empirical hypothesis. It does not itself incorporate sociological analysis, which must instead be the task of a sociology of knowledge itself following the same critical rationalist principles, which would seek to discover the social conditions of the emergence and maintenance of specific beliefs in specific societies. The separation of ideology-criticism and the sociology of knowledge is the consistent outcome of critical rationalism's insistence on the principle that the social origin of a belief is not relevant to the question of its cognitive validity. (Cf. Albert, 1968a, pp. 80ff.).

Nevertheless, Richard Münch argues that ideology-criticism on critical rationalist principles can contribute to human emancipation, thus rendering the pretensions of critical theory nugatory. Although theories of justice, which might claim to prove that an egalitarian distribution of life-chances or a particular form of political organization was inherently just, are deemed to be unfalsifiable because they have the form of 'empty formulae' or tautology (*Leerformel*), the value of equality can be treated as an orientation point, arrived at by an extra-scientific decision, for scientific research. Thus 'ideological' theories of justice and equality can be replaced by scientific research into the conditions under which life-chances could be extended and distributed more equally. Given the extra-scientific value of equality, critical reason can contribute to its realization by inquiring into the technical question of how it might be brought about, just as nomological knowledge in all sciences can be transformed into technical recommendations which can be followed in order to facilitate human emancipation from natural or social constraints (Münch, 1973, pp. 159, 166).

Among the ideologies that are condemned as such by this critical rationalist theory of ideology is critical theory itself. Precisely in its refusal to accept that emancipatory knowledge in the context of social action can take the same technically transformable nomological form as it does in the natural sciences, critical theory becomes subject to the charge of perpetrating ideology. Measured against the standards of critical rationalism, both the epistemological and the social propositions of critical theory must be allocated to the category of empty formulae, whose truth value cannot be decided. In addition, in its insistence that the forms of human knowledge must be understood in terms of the interests which underlie them, critical theory is seen as offending against the principle of the irrelevance of the origin of knowledge to its validity, and is condemned as an ideological threat to the 'public character of scientific method', allowing social interests to intrude into the procedures by which hypotheses are advanced and freely criticized.

CHAPTER 12

THE CONCEPT OF IDEOLOGY IN CRITICAL THEORY

The successive theories of ideology advanced by representatives of critical theory must be seen in relation to their interpretation of Marx and their assessment of the applicability of Marx's categories to the understanding of advanced industrial societies. As has already been remarked, the members of the early Frankfurt School accepted Marx's critique of political economy as essentially correct, and interpreted it in ideology-critical terms. They therefore took the emphasis away from the base-superstructure metaphor, which in Engels and in the Marxism of the Second International tended to generate the ontological thesis that all thought is dependent on its material substratum, for in this version, as Wellmer puts it,

> the concept of ideology loses the unique strategic significance which it had in Marx's theory, namely, that of a concept signifying a *false* consciousness which *in its falseness adequately expresses* and reflects a 'false' social reality (1977, p. 236).

In the same vein, Horkheimer (1967) immediately rejected Mannheim's 'total' concept of ideology and the project of the sociology of knowledge in general as necessitating an abandonment of the critical impulse of the critique of ideology.

Marx's critique of political economy derived its cogency from its relationship to the social force which had both a potential for transforming society by abolishing the capitalist mode of production and an interest in that abolition, the proletariat. It also depended, as the members of the Frankfurt School came to realize, on the continued validity of the basic categories of the critique, in particular the relatively clear separation of economics from politics, which allowed the 'laws' of capital accumulation to take their course independently from political intervention, and rendered capitalist economies susceptible to successive and uncontrollable economic crises. With the publication of the *Dialectic of Enlightenment*, Horkheimer and Adorno followed through the implications of the assumption that neither of these conditions any longer held valid. Capitalist economies had been transformed from within by the intervention of the state to control crisis by budgetary and fiscal manipulation and the management of investment, while at the same time the organization of the proletariat in trade unions and social democratic parties had resulted in higher living standards and the incorporation of the proletariat into an increasingly stable capitalism. In these new conditions, as Habermas puts it, "a critical theory of society can no longer be constructed in the exclusive form of a critique of political economy" (1971b, p. 101), and the critique of ideology, if it was to be continued, had to be transformed.

The reaction of Horkheimer, Adorno and Marcuse was to develop 'the critique of instrumental reason'. The assumption behind this was the view

> that with the emergence of organized capitalism a closed universe of 'instrumental reason' or 'one-dimensional rationality' was being created which not only threatens the chances of emancipatory political movements but also threatens the emancipatory impulses of the suppressed masses as such (Wellmer, 1977, p. 244).

Rather than the increased capacity of humanity to control and manage its dealings with natural forces generating an emancipatory potential, as Marx assumed, the critical theorists saw in this expansion of instrumental rationality a new form of repression and oppression, in which the consciousness of those who might otherwise be the agents of emancipation became increasingly reified and controlled. In the critique of instrumental reason, however, the concept of ideology is transformed. Marcuse argues thus:

> The relation between ideology and reality is a thoroughly historical relation, and as such determined by changes in society. In the Marxist view, 'ideology' refers to a consciousness which hurries on in advance of present reality, to the extent that it projects ideas (e.g. freedom, equality, happiness) which are made possible by social development but at the same time prevented by it from being realized. Incapable by itself of altering this situation, and controlled by social reality, ideological consciousness is a 'false consciousness', but as such it anticipates in idealistic form historical possibilities which are held in check by the reality of the present. This concept, however, seems not to be applicable to highly developed industrial society. This society has overcome ideology by translating it into the reality of its political institutions, its owner-occupied houses, atomic power stations, supermarkets, department stores, and psychiatrists' consulting-rooms (1967, p. 387).

In other words, instrumental reason, and the 'technocratic consciousness' which incorporates it, does not "express a projection of the 'good life'" (Habermas, 1971b, p. 111), which can act as a point of reference for the immanent criticism of ideologies, as the bourgeois ideals of the constitutional state and of free and equal exchange acted as points of reference for Marx's immanent criticism. Marcuse recognized the problem this posed for critical theory very acutely, for he saw that an ideology-critical critique of instrumental reason, according to which "the very concept of technical reason is perhaps ideological" (1972a, p. 223), will itself appear in the context of advanced industrial society as a regression to speculative utopianism, and to that extent itself "ideological" (1967, p. 387). The alternative which Marcuse, as well as Horkheimer and Adorno, imagined, must appear as a complete break with instrumental reason itself, so that emancipation was conceived not solely in terms of the transformation of social relations but also in terms of a completely new relationship to nature and to productive labour and its organization, a new science and a new technology.

In the judgement of Habermas, and following him Wellmer, this position is highly unsatisfactory: the critique of instrumental reason is too undifferentiated a concept with which to pursue the intentions of critical theory. In Wellmer's view, the eschatological quality which the idea of liberation takes on in the critique of instrumental reason is the other side of the coin to Marx's 'latent positivism'; for whereas Marx's exclusive concern with productive labour contained the danger that emancipation would come to be synonymous with the technical control of the systemic disequilibria to which capitalism was prone, the use of the concept of instrumental reason to refer to the basis for technical control of both 'external' and 'internal' nature led the members of the Frankfurt School to the oversimplified standpoint that technical rationality in both spheres had to be transcended for emancipation to be possible. Both in the case of Marx, at least in that aspect of his writing which exhibits this latent positivism, and in the case of Horkheimer, Adorno, and Marcuse, the mistake, according to Habermas, was to fail to distinguish adequately between the spheres of instrumental and communicative action, and hence between the technical and the practical. What had happened in the administered societies of managed capitalism and managed 'communism' was that the categories of instrumental action had supplanted those of communicative action in men's and women's understanding of the institutional framework in which they lived. With increasing state intervention in the running of the economy, and the growth of the welfare state, all questions of the improvement of living conditions had come to be seen as technical problems of social adminstration, rather than as ethical or practical political issues.

Thus for Habermas, what is important about technical reason

> is that it can also become a background ideology that penetrates into the consciousness of the depoliticized mass of the population, where it can take on legitimating power (1971b, p. 105).

The forms of capital utilization of capitalist societies, the distribution of income and wealth in those societies, and the forms of political decision-making, can no longer be legitimized by the ideology of equal exchange, since the state itself disrupts the exchange relationships of the market. Instead, the task of legitimation is taken over by appeal to the necessity for technical decisions to be taken by experts to meet the objective demands of the maintenance of economic growth, and hence the maintenance of supply of welfare benefits and security with which the support of the electorate may be purchased. In this sense, science and technology function in the place of earlier ideologies. But the concept of ideology itself has to change in the process. Firstly, the technocratic ideology is according to Habermas not a class ideology which serves to maintain the domination of one class and the subjection of another. It is

> more irresistible and farther-reaching than ideologies of the old type. For with the veiling of practical problems it not only justifies a *particular class's* interest in domination and represses *another class's* partial need for emancipation, but affects the human race's emancipatory interest as such (1971b, p. 111).

Secondly, the technocratic ideology is

> 'less ideological' than all previous ideologies. For it does not have the opaque force of a delusion which is only a pretence of the realization of interests (1971b, p. 111).

In other words,

> today it is no longer just the oppressed parts of society which with the aid of ideologies are pacified or fobbed off with lies. It is rather that the whole society is blinded by its own technocratic confession of faith – by an ideology, that is, that no longer directly lies, because science and technology really do make possible the realization of universal goals (such as the raising of living standards); but it is precisely in this way that this technocratic ideology is all the more effective in keeping people from an insight into their 'true' needs and interests (Kreckel, 1975, p. 96).

Ideology which appeals to science and technology is thus in one sense "less ideological", and on the other hand even more effective as ideology and "less vulnerable to reflection" (Habermas, 1971b, p. 111).

Habermas thereby assumes that the science and technology which make possible expanded control over 'external' natural forces and hence improved living standards are indeed 'indispensable as such'; what is ideological is not, as Marcuse supposed, the form of such scientific and technological knowledge itself, but its migration from the sphere of instrumental action, which is its proper place, to the sphere of communicative action, where it distorts the categories of practical reasoning and turns questions which should be open to political and ethical decision into ones which can be dealt with as problems of technical manipulation. Alongside the sociological aspect of the concept of ideology, therefore, which focusses on the legitimizing functions of technocratic consciousness, goes an epistemological critique; for the consequences of this new ideology in the theory of the social sciences is precisely to reinforce objectivist and and positivist tendencies which aim to make the study of society scientific in the same sense as the natural sciences. This is particularly clear in the doctrines of critical rationalism itself, which embraces not only a doctrine of the unity of scientific method, but also the social-theoretical implications of that doctrine, namely the technological application of social-scientific knowledge for purposes of social engineering (Popper, 1961, pp. 42ff.). Critical rationalism, and other 'positivist' versions of the unity of scientific method, are thus not merely mistaken theories of knowledge whose significance is limited to the specialized epistemological level of argument, but are also outgrowths from or aspects of technocratic ideology:

> The depoliticization of the mass of the population, which is legitimated through technocratic consciousness, is at the same time men's self-objectification in categories equally of both purposive-rational action and adaptive behaviour. The reified models of the sciences migrate into the sociocultural life-world and gain objective power over the latter's self-understanding. The ideological nucleus of this consciousness is the *elimination of the distinction between the practical and the technical.* It reflects, but does not objectively account for, the new constellation of a disempowered institutional framework and systems of purposive-rational action that have taken on a life of their own (Habermas, 1971b, pp. 112f.; cf. 1972, p. 316).

It follows that the critique of technocratic ideology must include the critique, as mistaken epistemology *and* as ideology, of positivist theories of the social sciences, among them critical rationalism.

The basis for such a critique of ideology cannot be purely immanent criticism, nor, as in Marx, immanent criticism combined with the emancipatory standpoint of the proletariat, since the ideology which has to be penetrated is not a class ideology, nor does Habermas have much faith in the emancipatory potential of the proletariat. Instead, the critique of technocratic ideology and its positivist forms of expression must be rooted in the conception of the practical and emancipatory interests themselves, and the notion of autonomy and responsibility (*Mündigkeit*) which the emancipatory interest anticipates. As Habermas puts it,

> the reflection that the new ideology calls for must penetrate beyond the level of particular historical class interests to disclose the fundamental interests of mankind as such, engaged in the process of self-constitution (1971b, p. 113).

This explains the strategic location of Habermas' more recent writings on the theory of language in his overall ideology-critical project; for technocratic ideology

> violates an interest grounded in one of the two fundamental conditions of our cultural existence: in language, or more precisely, in the form of socialization and individuation determined by communication in ordinary language. This interest extends to the maintenance of intersubjectivity of mutual understanding as well as to the creation of communication without domination. Technocratic consciousness makes this practical interest disappear behind the interest in the expansion of our power of technical control (1971b, p. 113).

The critique is thus supposed to take its force from an analysis of language, and from the assumption, which, it is claimed, is built into the use of ordinary language, of the possibility of unconstrained consensus and undistorted communication. These ideas act as the concept of 'truth' which functions as the contrast-concept to ideology in critical theory's current theory of ideology (Wellmer, 1977).

Unlike critical rationalism's theory of ideology, therefore, critical theory's theory of ideology does not directly contrast 'ideology' and 'science'. Enlightenment is not to proceed by replacing ideological consciousness with scientific consciousness, for outside the legitimate sphere of instrumental action scientific consciousness has, according to the critical theorist, itself become ideological. 'Ideology' is instead contrasted with a "reflective dissolution of false consciousness" (Wellmer, 1971, p. 72), which far from requiring the application of scientific method to the critique of social beliefs which legitimate inequality and domination, requires instead its containment.[48]

CHAPTER 13

THE CONCEPT OF IDEOLOGY IN SCIENTIFIC REALISM

The concept of ideology plays an important role in attempts to explicate an antipositivistic, realist naturalism which take as their point of reference the writings of Marx, and whose main doctrines have already been outlined in Section 8.3.2.[49] The significance of the concept of ideology in realist Marxism is both sociological and epistemological: on the one hand the term is used to refer to (mainly) class-based systems of belief which on realist principles are causally dependent on the material location of each class in a mode of production and which also play a part in the maintenance or transformation of social formations; on the other hand 'ideology' refers to forms of distorted or false consciousness (Keat and Urry, 1975, pp. 176ff.; Benton, 1977, pp. 161ff.; Bhaskar, 1979, pp. 80ff.). The distinctive feature of the epistemological aspect of realism's theory of ideology is that the falseness of ideology is measured against the 'truth' which is made accessible by realist science; 'ideology' is contrasted with 'science', while science is interpreted on realist lines.

As we have seen in presenting critical theorists' interpretations of Marx, it is possible to derive two different theories of ideology from Marx's writings. In the first, ideology-critical theory, 'sociological' and 'epistemological' aspects are tied together by the epistemological role played by the proletariat in providing the standpoint from which, not only may capitalist ideologies be immanently criticized, but the 'truth' of proletarian criticism may also be 'realized in practice'. In the second, which critical theory brands as exhibiting 'latent positivism', the theory of ideology becomes a universal sociology of knowledge, proposing that all thought, all consciousness, is causally dependent on its material basis. In the latter theory, the sociological and the epistemological elements of the concept of ideology are separated, and can only be related to each other externally; a further criterion is required to show why the consciousness of the proletariat is potentially true. It is this second theory that provides the basis for scientific realists' theories of ideology. On the one hand, the sociological aspect of the theory does become a generalized sociology of knowledge in which the manner in which knowledge is produced in a society, and its content, are directly related to the social relations of production (Keat and Urry, 1975, p. 177).

This is made even clearer by Benton, who argues that

ideologies, in their practical forms, are present in all social practices, including economic ones . . . The different class ideologies, then, have their material existence in the different practices and experiences of the opposed classes. The source of ideologies is therefore to be sought in the conditions of existence of the classes and their practices themselves (1977, pp. 177f.).

On the other hand, ideology is contrasted epistemologically with science, and regarded as false in relation to scientific truth. From this point, in which the sociological elements are definitely separated from the epistemological elements of the concept of ideology, it becomes highly problematic why the consciousness of one class should be any more true than that of any other. Rather than the truth of the proletariat's standpoint being built into the theory of ideology by its own presuppositions, this becomes a completely contingent question. As Keat and Urry have it, on the one hand Marx's own theories were supposed to be "in the interests of the class of wage-labourers" (1975, p. 178), on the other hand they were also supposed to be non-ideological, that is, scientific. This proves the realist thesis that both true and false beliefs may be given causal explanation; conceptually, however, truth and class origin are independent.

In order to give some foundation to the epistemological aspect of realism's theory of ideology, a theory of knowledge is required which gives some content to the distinction between science and ideology, and explains the basis of the latter's falsity and the former's potential truth. Keat and Urry, as we have seen, have little to offer in the way of such a theory. Realist science is distinguished from ideology by the assumption that realist science is capable of discerning the real but unobservable causal mechanisms and structures of capitalist society, whereas ideology is unable to discern these mechanisms and structures accurately, because "capitalist society appears to its members in a systematically misleading form" (1975, p. 179). Yet as Keat and Urry themselves recognize, there is a problem in this formulation.

For Marx the correct scientific theory enables one to reject the ways in which social phenomena are normally conceptualized and described. But if we are able to redescribe what takes place in a society in terms of a new scientific theory, then how can that theory itself be defended against those who reject the view that the existing descriptions are themselves distorted? What is it about one theory-dependent description that makes it preferable to another? (1975, p. 180.)

Unfortunately, although Keat and Urry spot the problem, they have no answer to it, and their attempt to construct a realist theory of ideology is weakened as a result.

Benton, on the other hand, pays more attention to the problem of justifying the distinction between science and ideology, for he identifies as a major problem area for a materialist and realist theory of knowledge:

If science and ideology are to be theorised as distinct historical realities, then theoretical criteria need to be established for distinguishing them (1977, pp. 173f.).

He pursues this problem by mentioning a number of possible criteria which he finds in the writings of Althusser. He rejects a number of such approaches. He rejects Althusser's suggestion that there is a 'mechanism of the knowledge effect' which ensures that the 'object of knowledge' of a science is also the 'real object', whereas the 'object' of an ideology is not, on the grounds that this merely restates the problem, and raises the insoluble Kantian problem of the 'thing-in-itself'. He rejects the proposal that ideology

is distinguished from a science not by its falsity, for it can be coherent and logical (for instance, theology), but by the fact that the practico-social predominates in it over the theoretical, over knowledge (Althusser, 1969, p. 251),

on the grounds that this criterion would make the distinction between science and ideology one of degree, and that in any case it cannot be applied, since it is not clear how 'practico-social functions' are to be measured. Other proposals are briefly dismissed, including the denial of the very possibility of a general distinction between science and ideology, which would collapse into relativism. Finally Benton mentions one criterion which he believes does promise more fruitfully to make the distinction required.

This is the distinction between science and ideology in terms of the character of their internal unity. The problematic of a theoretical ideology is said to be 'closed' whereas that of a science is 'open'. The 'closure' of ideological problematics is, further, connected with the type of relationship they have with non-theoretical practices. The exigencies and demands of the other social practices present to theoretical ideology 'solutions' for which the theoretical ideology must produce, in a theoretical form, the appropriate

> 'problem'. In other words, the concepts of a theoretical ideology will be structured around a problem, whose solution is already given in the form of a demand imposed by extra-theoretical practices . . .
>
> By implication, the 'openness' of the problematic of a science consists in its 'solutions' not being predetermined by the structure of its theoretical problems and in its problems not being set by extra-theoretical requirements and interests: the 'objectivity' of science consists precisely in these differences between it and theoretical ideologies (1977, pp. 188f.).

Benton goes on to illustrate this distinction with examples drawn from the natural sciences. 'Medieval cosmology' is characterized as a theoretical ideology, because although it allowed a certain degree of internal development its solutions were fixed by its relationship to the demands of Catholic theology. The scientific revolution of the sixteenth and seventeenth centuries instituted an open problematic by becoming conceptually autonomous from those demands. This does not mean, Benton stresses, that science did not also serve social interests; the new natural science served, and Benton implies it continues to serve, the interests of the capitalist class, then in its struggle against the feudal order, now in its defence of its ruling positions. But that this is so is quite independent of its status as science, which is determined by its openness to any solutions to its problems which research may throw up. True to the principles of realist naturalism, Benton assumes that precisely the same distinction between science and ideology can be made in the realm of thought about society as he sketches out in relation to the study of nature.

In sum, it seems as if the concept of ideology in realist Marxism is carrying a considerable weight. On the one hand it plays a part in a comprehensive sociological theory of knowledge, which, in a way reminiscent of Mannheim (who is seldom referred to by realist writers), aims to relate all thought to the social standpoint of those who think it. On the other hand, it is a key component of a separate epistemological theory of knowledge which, like critical rationalism, raises the standards of science to the status of sole criterion of truth, and judges the falsity or distortedness of ideology from the point of view of that standard. Only the second aspect of this theory of ideology has critical potential, and in this case 'enlightenment' must come from the critical confrontation of inadequate ideological consciousness with the findings of science, both natural and social, about the true structure of 'reality'.

CHAPTER 14

DISCUSSION: POSITIVISM, ANTIPOSITIVISM AND IDEOLOGY

The relationships between the theories of ideology advanced by critical rationalism, critical theory, and scientific realism may be briefly summarized. Critical rationalism's essentially epistemological theory of ideology, which has the consequence that the doctrines of critical theory also come under the rubric of ideology and are criticized as such, itself depends on the universal applicability of the methods of critical reason as critical rationalists understand them. This is precisely what critical theorists dispute, since for them the hermeneutic and historical nature of social and cultural formations preclude the automatic assumption that they too are governed by universal laws which the methods of critical reason will be best suited to discover. In the discussion to Part III (Chapter 9) the weight of epistemological argument was judged to favour the critical theorists, and if this is so, the force of critical rationalism's theory of ideology would be removed.

On the other hand, critical theory's concept of ideology, which operates on both an epistemological and a sociological level, on the assumption that a critical theory of knowledge cannot hope to be fruitful as pure philosophy but must take the form of a theory of society, itself depends, in the most advanced formulation proposed by Habermas, on the validity of the distinction between the spheres of instrumental and communicative action, between sensory and communicative experience, and on the viability of the theory of the different human interests which come into play in each sphere. On the basis of this theory of ideology, what appears to critical rationalism as enlightenment appears to critical theory as ideological, for the notion that greater freedom and equality can be pursued as a technical project by social engineers using well-tested nomological theories of social development is exactly a form of the technocratic consciousness which the critical theorist identifies as the new ideology of administered highly developed industrial society, even when the technocrats represent themselves as Marxists pursuing the goal of the construction of a communist society. Such a notion requires the division of society into planners and planned, and thus maintains the asymmetrical relationships and distorted communication which are a sign of the degeneration of the difference between instrumental and communicative action. And yet it is the theory of interests built up by reflection on the methodical pursuit of knowledge in these different spheres which to the critical rationalist appears as the ideological importation of extra-scientific interests into science, which can only threaten its autonomy and act as a hindrance in the search for truth.

From the point of view of critical theory, the theory of ideology advanced by scientific realists must appear as little more than a variant of that proposed by critical rationalists. It is, like the critical rationalist theory, a 'scientistic' concept of ideology, which aims to distinguish between ideological falsity and enlightening truth with the methods of science, which will be the same in the study of society as in the study of nature. It also sees enlightenment as emerging through the scientific criticism of uncritically accepted understandings of both society and nature, and in one version at least conceives of the application of the fruits of this enlightenment as a technological project (Keat and Urry, 1975, p. 195). And in Benton's formulation of the distinction between science and ideology, the terminological coincidence with critical rationalism is so close that the passage quoted above could almost have been written by Popper himself. The stress on the 'openness' of science as compared with the 'closure' of ideology, which is made possible by the autonomy of scientific problem formation and problem solving from extra-scientific interests, appears indistinguishable from the demarcation advocated by critical rationalism. The realist separation of the class origin of a system of belief from its truth on scientific criteria can also be viewed as a version of the doctrine, central to critical rationalism, that the source of a theory is irrelevant to its validity. It would appear, therefore, as if the attempt to interpret Marxism on the naturalist principles of scientific realism leads its advocates to a theory of ideology which is not substantially different from that of critical rationalism, despite differences in intention, tone, and rhetoric. From the standpoint of critical theory, this would not be a surprising outcome, but would be expected of any attempt to interpret Marxism as an objectivizing science. As Habermas wrote of dialectical materialism:

As long as historical materialism no longer saw itself as involved in the objective crisis complex, as soon as it understood its critique exclusively as a positive science and the dialectic objectively as the law of the world, then the ideological character of consciousness had to take on a metaphysical quality. Spirit then was considered simply and always as ideology, and this included socialism. Within this superficial understanding, the correct ideology was distinguished from the false solely according to the criteria of a realist theory of knowledge (1974, p. 238).

This judgement applies even when 'realist science' is substituted for 'positive science'; and even though realists avoid speaking of 'true ideology', calling it 'science' instead, they still have to assimilate both true belief and false to a general causal theory of knowledge, as did dialectical materialism. The objectivism shared by critical rationalism and scientific realism (and, for that matter, dialectical materialism) generates the same kind of 'scientistic' theory of ideology. And to the extent that scientific realism too fails to recognize the significance of the distinction between instrumental and communicative action, and in its objectivism fails to reflect on the meaning of the pursuit of knowledge in these different spheres, it too would be subject to the judgement that it had been unable to detach itself from the modern 'ideology' of science and technology.

Reciprocal charges of being 'ideological' are an obstacle to mutual understanding and to the achievement of agreement, as the course of the *Positivismusstreit* demonstrates. Yet the lack of mutual understanding between critical rationalists and critical theorists seems to be difficult to reduce. Even the attempt to formulate each in comparable epistemological terms runs up against the barrier that the aim of each in pursuing the theory of science is far removed from the other. The aim of the critical rationalist is to specify the conditions, both methodological and social, which will facilitate the progress of nomological, or empirical-analytic, science; in a sense, critical rationalism can be considered as itself a technology of scientific progress. The aim of the critical theorist is both hermeneutic and emancipatory:

Theory of science in the medium of critical reflection is thus directed not just toward the maximization of the progress of scientific knowledge but toward the idea of successful communicative mutual understanding concerning the social system of interaction called 'science' (Schnädelbach, 1972, p. 112).

The critical theorist is concerned with "what role one accords science in the life of the mind and ultimately in reality", as Adorno put it, continuing:

Dialectics remains intransigent in the dispute since it believes that it continues to reflect beyond the point at which its opponents break off, namely before the unquestioned authority of the institution of science (Adorno *et al.*, 1976, p. 67).

This continued reflection is reflection in both the Kantian and the Hegelian senses; it is epistemological reflection on the conditions of possible knowledge, and it is also critical reflection on humanly produced constraints which are in principle removable by liberating practice. These come together in the theory of science, since according to critical theorists one of the major humanly produced constraints which now has to be overcome is precisely "the unquestioned authority of the institution of science", which performs ideological functions in advanced industrial societies. One will only fully accept the force of critical theory's epistemological arguments if one also recognizes the validity of the emancipatory intention, and vice versa. The critical rationalist accepts neither.

The position of the 'realist Marxist' is somewhat different. Even if the realist theory of science, as it is at present propounded, is an objectivist rather than a reflective understanding of science, and even if its transference in a naturalistic form to the theory of the social sciences is evidence that it is unable by itself to draw distinctions which would enable it to break decisively with the technocratic ideology of modern society, as the critical theorist would see it, the realist Marxist nonetheless shares with the critical theorist one common principle which should make possible mutual understanding and even prepare the way for agreement, namely that theory should contribute to emancipatory practice. Unlike the critical rationalist, both the critical theorist and the realist Marxist are guided in their reasoning by an anticipation of a classless society or, in more abstract terms, of a total social subject (Wellmer, 1971, p. 135), and intend that social theory should develop in relation to practical activity directed towards emancipation from the constraints which hamper the achievement of that anticipated state of affairs. In their 'positivistic' misunderstanding of the form of social theory that would be necessary to pursue that intention, realist Marxists are in somewhat the same position, so the critical theorist might judge, as Marx himself when, stimulated by Victorian enthusiasm for scientific progress, he misinterpreted his critical social theory as a natural science of society and gave expression to the 'latent positivism' of historical materialism. The acceptance by the realist Marxist of the emancipatory intention could make possible an understanding of critical theory's critique of positivism and of scientific realism's own objectivism, and enable the realist Marxist to undertake a reinterpretation of the realist theory of science with a view to identifying its critical potential.

Yet, if the argument of Chapter 9 is correct, this critical potential cannot lie in the direction of a realist theory of the social sciences, since realism's naturalism is as misguided as the 'positivist' version it opposes. The strength of scientific realism, as was suggested in Chapter 5, must lie where it was originally developed, in theoretical understanding of the natural sciences. Realist theories of science, by reflecting developments in the methodological procedures of natural sciences, may present a coherent and systematic alternative to traditional theories of science based on deductivist premises, and may thus constitute a reasoned challenge to claims to universality on the part of deductivism. Initially this challenge appears to have significance only within the philosophy of the natural sciences as this is pursued within academic philosophy. Now, however, in the light of critical theory's concept of ideology, it can be seen to have wider significance too. For by demonstrating that alternatives to deductivist conceptions of natural scientific knowledge cannot only be imagined but have indeed been realized in practice, it enables critical theory to extend its own critique of the role of the natural sciences in advanced industrial societies.

As we have seen, Habermas is reluctant to take the path pointed out by other critical theorists such as Marcuse and Sohn-Rethel, which would lead to a critique of the existing forms of the natural sciences as ideological and thus to speculation about an alternative form of natural science, because he believes that the demands of technical applicability rule out the possibility of such an alternative. Habermas goes so far as to suggest that such reflection is dangerous, and that an objectivist self-understanding on the part of natural scientists protects them from such distortions as fascism's 'national physics' and Stalinism's Soviet genetics (1972, p. 315). The ideological role of science and technology in advanced industrial societies is therefore restricted to the extension of their methods, reinforced by their objectivist interpretation, to the sphere of society; the theory of ideology is reformulated to become a critique of technocratic conceptions of political decision-making and social administration on the one hand, and of the positivistic social sciences which these require on the other. In the light of the possibilities presented by the realist theory of science, this restriction may turn out to be itself over-cautious. So long as scientific realism is not taken in its essentialist, objectivist form to be specifying the essence of science as such, but rather, as has been suggested here, is interpreted as a descriptive methodology appropriate to a particular phase in the development of certain natural sciences, the realist theory provides evidence of precisely that which Habermas denies, namely the possibility of the emergence of alternative methodological categories under which knowledge in the natural sciences can be organized. If this possibility is a real one, if the basic categories of scientific method are still subject to historical change, then the potential is available for a return to the theme of the earlier writers of the Frankfurt School, and for a reconsideration of the proposal, in Marcuse's terms, that "the very concept of technical reason", not as such, but as it is embodied in existing categories of scientific method in present-day industrial societies, is 'ideological'.

The problem can be seen as one of identifying an alternative concept of scientific progress in human knowledge of nature, more explicitly related to the emancipatory interest of reason. According to the deductivist principles of the nomological conception of scientific knowledge, the idea of deductive unity acts as a standard against which progress can be conceptualized and measured; in the terms of Kant, deductive unity is an Idea of reason. Such Ideas

> are not derived from nature; on the contrary, we interrogate nature in accordance with these ideas, and consider our knowledge as defective so long as it is not adequate to them (1929, p. 534).

Ideas of reason, which regulate the process of acquiring knowledge and function as a measure of progress, will however appear arbitrary unless they can be given some foundation or justification. Kant himself wanted to provide a foundation which would be valid for any consciousness whatever, and, although he conceded that even in his own terms he could not demonstrate the objectivity of the ideas of reason by transcendental deduction, as he claimed to be able to do in the case of the categories of the understanding, he did hope to root their objectivity in the natural tendency of reason to seek the ultimate limits of its applicability, which would thus provide a basis for the idea of striving towards a systematic unification of all knowledge (1929, pp. 549ff.). The critical theorist, as we have seen, remains dissatisfied with this form of argument, and instead hopes to find a foundation for the basic categories of knowledge in the inescapable practical situation of the human species. According to Habermas, the need to come to terms with non-human nature generates the search for reliable rules for the manipulation or evasion of natural forces in the system of instrumental action, rules which will be the more reliable the more universally valid and the more systematically interrelated they become. Systematic, and in the context of empirical-analytic nomological science this means deductive, unity is an idea which guides the progressive fulfilment of the inherent human interest in technical control over objectified nature, and this interest itself is one component, in the context of instrumental action, of the equally inherent human interest in emancipation, for technical control of nature will emancipate human beings, so far as is possible, from subjection to unknown and uncontrollable natural forces, and enable them to produce the material means which can be the basis for the construction of a more equal and just society. Thus Habermas' reformulation of the Kantian problem and his more pragmatist solution ends by justifying, at least in the context of the natural sciences, precisely the same conception of scientific progress; progress towards deductive unity will best serve the technical, and hence the emancipatory interest.

Scientific realism, as we have seen, attempts to demonstrate that, in the light of more recent developments in scientific method, deductive unity is an inappropriate and unattainable ideal for scientific knowledge, unattainable, not in the Kantian sense of being a limiting concept which can be imagined but not reached, but in the sense of being an ideal which does not provide an intelligible guide which could regulate scientific progress. This argument may well be overdrawn. For present purposes, however, what is significant is that realist theories of science suggest that deductive unity is not the only possible regulative idea which could guide scientific progress, and that others could be imagined which might provide realizable maxims for scientific research. If this is so, then it can no longer automatically be assumed that deductive unity is the best regulative idea to serve the human interest in emancipation in relation to nature. It may instead be, as Marcuse and Sohn-Rethel suggest, that deductive unity and its associated categories of scientific method, having arisen under particular historical circumstances and being associated with a particular form of the division of labour incorporating domination, are ideas the pursuit of which serves not universal interests in emancipation but particular interests in domination.

Deductive unity appears to provide an ideal for scientific knowledge and a goal towards which research may be directed which are internal to science itself, conceived as an autonomous sphere of human activity (cf. Heller, 1979). It appears to provide a conception of truth independent of the potential applicability of scientific knowledge, and seems to legitimate the separation of science and technology, or of basic and applied research. It appears to provide a touchstone for scientific progress which pays no attention to how this cumulative body of knowledge may be used in practice. Since the basic categories of scientific method are derived from the system of instrumental action, the results of this scientific progress are in fact technologically applicable, but the internal ideal of deductive unity makes this appear a by-product, which is of no concern to the scientist. The truth about nature must be pursued, it seems, without regard to the possible or actual use of scientific knowledge by others; application is the responsibility of industrial companies, politicians, even the electorate, of which scientists in their private capacity are members, but not of science itself. At this point the abstract philosophical argument which has been followed up till now begins to connect with contemporary political argument in many advanced industrial societies. For it has been held, by many scientists and non-scientists whose writings and pronouncements it cannot be the task of the present work to examine, that this conception of the autonomy of science is in fact an abdication of responsibility, and amounts to a concealed or tacit approval for the use of scientific knowledge for suppression and domination and the profit of the few, rather than the benefit of all (see Nowotny and Rose, 1979). Given the importance of science in modern technology, which has penetrated all aspects of life in industrial societies and elsewhere, this is one of the most fundamental political arguments of the modern world. But if it is to be conducted in rational terms within the scientific community itself, and not merely in confrontation in the environs of missile bases and nuclear power plants, the basis for the objectivity of the ideal of deductive unity, which provides the cornerstone of the argument for the autonomy and internal goal-setting of science, must itself be tested. If its objectivity, its universality and its necessity, cannot be grounded, as it seems on the basis of earlier argument that it cannot, then the case for the 'autonomy' of science is weakened, and the case for an alternative conception of scientific progress bound to 'external' goals of the satisfaction of real human needs is strengthened.[50] The 'unity' of science would not, on such a conception, be found in its internal deductive systematization, but in the common direction of scientific research towards human emancipation. If such a conception of scientific progress can indeed be formulated, then it would be appropriate to extend the critique of 'positivism', as the earlier members of the Frankfurt School tried to do, to the theory of the natural sciences, and to argue that 'positivistic' theories of the natural sciences perform ideological functions in modern societies and must be exposed to criticism.

NOTES

[1] The exception to this is Halfpenny's account of central tenets, which claims to be no more than an aid to preliminary discrimination of different models of sociological explanation, and does not claim "that these are the critical features that demarcate positivism from any other school of philosophy" (1976, p. 21). This disclaimer does not buy as much immunity from criticism of the 'central tenets approach' as the author would like.

[2] For the hermeneutic concept of tradition and its relation to language, see Gadamer (1975). Williams (1976), while not explicitly using the terminology of hermeneutics, is concerned with similar problems. For an example of the unravelling of the historical development of a specific word, see Macpherson (1966).

[3] The same issues are at stake in the reception by critical rationalism of Kuhn (1962). See Lakatos and Musgrave (eds.) (1970).

[4] This is not to say that there are not other important differences between critical rationalism and logical positivism; but they can be treated as subsidiary to this essential difference. Other differences, in particular Popper's embracement of 'metaphysical realism', will be treated below.

[5] Compare the distinction between 'thesis-pragmatism' and 'methodological pragmatism' made by Rescher (1977, pp. 37–80).

[6] For further discussion of the later epistemology of critical theory, see Held (1980), Kortian (1980), McCarthy (1978).

[7] The systematic and the historical components of Sohn-Rethel's analysis cannot, however, be easily taken apart. Münch (1973, pp. 106ff.) therefore misses the force of their combination when he argues against Sohn-Rethel that the fact that mechanistic thought arose under conditions of commodity-exchange says nothing about the subsequent relationship between that structure of thought and that form of social organization.

[8] It is not necessary here to enter on the question of what the end points in these regresses might be or, in particular, on the implications of realism for the nature of the 'ultimate' constitution of the world. For a discussion of realism, the regress of micro-explanation, and quantum theory, see Madden and Sachs (1972).

[9] Cf. Gaston Bachelard's conception of scientific instruments and 'phenomenotechnics', as discussed by Lecourt (1975, pp. 137ff.) and Bhaskar (1975).

[10] Zilsel's work was in some ways an advance on that of, say, Hessen (1971) or Grossmann (1935), in concentrating on the emergence of the category of laws of nature, which other writers take for granted.

[11] See, for example, the remark of Max Horkheimer (1972, p. 154): "Under present-day methods of production, science is more interested in the results of abstraction than in the theoretical reconstruction of the whole". Cf. Bohm (1971); see also the papers produced for the project 'Alternativen in der Wissenschaft' by members of the Max

Planck Institut zur Erforschung der Lebensbedingungen der wissenschaftlich-technischen Welt, Starnberg, e.g. Böhme *et al.* (1972, 1976).

[12] This is a summary of the argument of Chapter 2 above, resting on Popper (1968). After the polemical phase of the Positivismusstreit, this 'decisionist' and 'methodological conventionalist' standpoint was at last again formulated clearly by Albert (1968a, Chapters I–III).

[13] Cf. Connerton (ed.) (1976, pp. 17ff.). At the same time, the term 'reflection' also has two corresponding meanings: reflection on the conditions of the faculties of the knowing subject, and reflection on humanly produced constraints. As Habermas (1973a, p. 182) points out, "the traditional use of the term 'reflexion', which goes back to German Idealism, covers (and confuses) these two senses". Considerable confusion has been caused by failure to separate out these two meanings: see, for example Lobkowicz's (1972, p. 207) ironic puzzlement as to what reflection refers to. It would be advantageous to introduce a terminological convention to resolve the confusion, but this would be unlikely to succeed. Apart from an occasional attempt to use 'reflection' for the first sense and 'self-reflection' for the second, I therefore leave it to the hermeneutic powers of the reader to decide from the context which sense is predominant. They are, in any case, by no means unrelated.

[14] Cf. Habermas and Luhmann (1971, pp. 206ff.). More recently Habermas has suggested that in communicative action it is less important that experiences are constituted than that utterances are generated; he therefore adopts for the analysis of communicative action, or of speech acts in particular, a model of surface and deep structures; cf. Habermas (1976a, pp. 202f.). It is not possible to enter into discussion of these later developments in Habermas' thought, but they do not in any case affect the basic structure of the critique of positivism and of the doctrine of the unity of science.

[15] Cf. Popper's (1969, p. 113) similar conclusion on the natural sciences: "For instrumental purposes of practical application, a theory may continue to be used even after its refutation, within the limits of its applicability".

[16] This formulation, deriving from Kreckel (1972, p. 36) and Pilot (1976, p. 275), also provides a way of conceptualizing the problem discussed in the previous part of this work, for what was put into question by Marcuse and Sohn-Rethel was whether the existing methodology of the natural sciences really does conform to the emancipatory interest in reason.

[17] This thesis clearly bears a strong resemblance to the basic principles of ethnomethodology, at least in its earlier forms, which took the reflexivity of indexical expressions to be the fundamental feature of social interaction with which sociology had to, but had not yet, come to terms; see Garfinkel (1967) and Coulter (1971).

[18] This argument, developed against the claims for hermeneutics put forward by Gadamer, has given rise to considerable debate between critical theorists and hermeneuticians, which cannot be entered into in detail here. See Apel *et al.* (1971), and commentaries on the debate by Misgeld (1977), Bleicher (1980) and Thompson (1981).

[19] This critique of positivistic objectivism in Dilthey and historicist analysis of understanding is shared by Gadamer, and partly based on his work. Gadamer's conception of hermeneutics would not be subject to the critique of latent positivism, whatever other objections Habermas might have to the 'claim of universality' of hermeneutics. Cf. Gadamer (1975, pp. 192ff.); Habermas (1970a, pp. 253ff.).

[20] Habermas' parallel criticism of Freud's latent positivism will not be discussed in the

present work. This may be thought somewhat unfortunate, since one representative of the realist theory of science, Russell Keat, has chosen to debate with Habermas on the very ground of his interpretation of Freud (Keat, 1981). However, despite its intrinsic interest, Keat's criticism of Habermas in this context does not seem to depend on the former's adherence to realism, and in some aspects seems similar to that of such writers as Giegel (1971) who do not embrace a realist theory of science. As is argued later, Keat also misunderstands Habermas' epistemological position at crucial points, and this weakens the force of his criticism.

21 The problems of Marx-interpretation are well-known, and a discourse on them here would be out of place. Rather than having to decide on the 'authentic' texts, so that others may be seen as 'deviant', it is probably more fruitful to identify points at which incompatible approaches compete within the text, which set problems for later theorists, problems which are reduced in significance by making them problems of Marx-interpretation alone, (Cf. Henry (1976) Vol. I, Introduction: 'La theorie des textes'.) Thus at many points in *Capital*, and, for example, in the fashionable missing chapter of *Capital*, Vol. I, the ideology-critical and the 'positivistic' approaches continue to co-exist: cf. Marx (1976, Appendix).

22 Halfpenny (1976, p. 31f.) designates the combination of positivism and anti-naturalism as 'impossible', because he has already defined positivism in terms of the central tenet of the unity of science. This terminological decision ruled out consideration in his thesis of the issues raised by antipositivists in the present sense.

23 Giedymin (and, following him, Halfpenny) does briefly mention this possibility, and indeed identifies one antipositivist and antinaturalist writer, namely von Wright. This appears to be a rather curious conclusion, for, according to Giedymin, von Wright himself identified positivism with methodological naturalism (1975, p. 284), hence his opposition to one must be opposition to both. The two dichotomies are not separated, and little remains that is specifically both antipositivist and antinaturalist in von Wright's (1971) approach.

24 The textbook on methodology from a symbolic interactionist standpoint by Denzin (1970) also represents a deductivist conception of science. Glaser and Strauss's (1967) proposals for a symbolic interactionist methodology are a regression to an even more positivist, inductivist, standpoint.

25 Halfpenny provides helpful diagrammatic formulations of these alternative interpretations of Durkheim, taking the theory of suicide as his example. He leans towards the realist interpretation (1976, pp. 158–163).

26 This final point seems to rest on a confusion on the part of Keat and Urry concerning 'essentialism'. Durkheim's use of the term 'essence' does not seem to be a use which leaves him open to a Popperian critique; he might mean that definitions of this kind are descriptions of 'real essences', which Harré and Bhaskar see as an essential component of scientific knowledge but which Keat and Urry (1975, p. 48) see as inessential, or he might mean that scientific investigation must begin from an initial delimitation of a topic in terms of a 'nominal essence'; cf. Locke (1959, ii, p. 27).

27 It is of course impossible to enter here into the long and tortuous debate concerning the relationship between explanation in terms of causes and explanation in terms of reasons. Only the realist response to the positivism underlying the antinaturalist arguments is here relevant.

28 Bhaskar (1979, pp. 169ff.) also provides a detailed critique of Winch from the stand-

point of the former's 'transcendental realism'. In this he uses arguments, designed to show that Winch's antinaturalism is based on an untenable (positivist) theory of natural science (p. 173), which parallel those of Keat and Urry. He attempts greater systematization, however, and argues that significant aspects of the thought of such hermeneutical writers as Winch are dependent either on pure versions of positivist theories, or on modifications of them, formed by inversion or negation, displacement, transformation and condensation (p. 169). Bhaskar's discussion of Winch raises issues which go far beyond the immediate concerns of the present text, and the basic structure of realist critique of positivist antinaturalism stands out more clearly in Keat's earlier article.

29 Harré and Secord draw explicitly on the work of philosophers who have analyzed the categories of human action, such as P. F. Strawson and S. Hampshire, as well as on that of Erving Goffman.

30 There is a clear similarity between this conception of negotiation of accounts and the practical process of hermeneutic interpretation as discussed by Habermas, following Dilthey and Gadamer. This will be taken up in Chapter 9.

31 This is Keat and Urry's (1975, p. 100) account of Marx's concept of commodity-fetishism, which is here reported rather than presented as accurate.

32 Keat and Urry (1975), while claiming to eschew the vocabulary of the "visual metaphor of appearance-reality" (p. 100), still see Marx as providing causal explanations of beliefs in terms of generative mechanisms (pp. 192f.).

33 The thesis that reasons can in general be causes of actions is contradicted by Harré and Secord (1972, pp. 160f.); they suggest a distinction between the conception of a 'passive agent', who merely receives reasons for action as a condition of action, and that of an 'active agent', who actively makes decisions in a normative or justificatory context and presents a reason in that context. The former case could be considered as causal, the latter not.

34 Keat and Urry (1975, pp. 181ff.) do discuss Lukács' solution to this problem, that the truth of the Marxist theory derives from the universal interests of the proletariat, but their rather sceptical comments on Lukács suggest that they do not themselves find this solution satisfactory.

35 It is difficult to see how this differs from Popper's notion of an increase in explanatory power, despite the realist critique of positivism.

36 Bhaskar (1979, p. 63) is an exception to this; his attempt to make the pre-constituted, 'concept-dependent' nature of social reality consistent with realism strains at the limits of his naturalism; cf. Benton (1981).

37 The recognition of this problem has led some Marxist writers to a position which appears to embrace a thoroughgoing relativism. Colletti (1972, pp. 234ff.), for example, realizing that capitalism is described differently in the concepts of political economy from how it is described in the concepts of Marxism, proposes that "there are two realities in capitalism: the reality expressed by Marx, and the reality expressed by the writers he criticizes". Beyond suggesting that the former reality is experienced "from the viewpoint of the working class" and the latter from the viewpoint of capital, there is no convincing argument advanced by Colletti which escapes this 'Mannheimian' relativism.

38 Both Marx and the critical theorists make use of the category of Naturwüchsigkeit to refer to processes of history and organizational forms of society which take a

quasi-natural form, obeying quasi-natural laws of their own, in contrast to a potential for social forms which are the result of the free and collective decision of the participants. See, for example, Marx and Engels (1968, p. 45), where 'naturwüchsig' is translated as 'natural' or 'naturally'.

39 Keat and Urry do not present any further analysis of this special ontological category.

40 Benton (1981) has criticized Bhaskar for giving far too much away to antinaturalism, and has attempted to revise Bhaskar's version of realism to avoid the need to postulate such limits to naturalism.

41 For a discussion of how these conditions may be analyzed, see Habermas (1970b, 1973c, 1976a) and Habermas and Luhmann (1971, pp. 101–141). For a critical summary and discussion of Habermas' more recent work, see McCarthy (1972, 1978). A more detailed consideration of the issues raised by this work would go beyond the bounds set for the present essay; my concern is not to show that critical theory has sufficiently solved the problems it throws up, but rather to use critical theory as a measure in relation to which the inadequacies of a realist theory of social science may be discerned, even though each in its way has much to contribute to the critique of positivism.

42 'Herrschaft' is better translated as 'domination' than as either 'power' as J. J. Shapiro has it in Habermas (1972) or as 'authority' as G. Flöistad has it in Habermas (1966). For a discussion of the concept of 'Herrschaft' in critical theory, see Geuss (1981), esp. pp. 16ff.

43 For a discussion of the difficulties of distinguishing between 'true' and 'false' consensus, see Habermas (1976b, pp. 111ff.) and McCarthy (1972, 1978).

44 I confess to finding Bhaskar's (1980) attempt to transform transcendental realism into an emancipatory critical naturalism which is not objectivist very difficult to understand. Two interpretations seem possible: either that Bhaskar has moved so far away from a realist theory of social science that it would be best to abandon any pretence to naturalism (cf. Benton, 1981); or that his conception of emancipation is to be equated with knowledge of necessity. The latter interpretation is supported by his insistence that "for emancipation to be possible, knowable emergent laws must operate" (1980, p. 28). The challenge to this is the same as the challenge to critical rationalism: to present any plausible candidate for such a law.

45 It is therefore a complete misunderstanding on the part of Scott (1978) to argue that Habermas' critical theory defends a realist epistemology, or that Habermas "is grappling with the very difficult epistemological problems of a 'realist' social science".

46 This statement should not, however, lead us to ignore the fact that communicative action is itself dependent on a material base, which forms a medium through which communication may flow, and that developments in the control of natural forces may change the nature of that material base, thus altering the possibilities of communicative action. This is a rather abstract way of conceptualizing the obvious development of communications media, which have developed from the basic use of sound and sight to carry ordinary-language speech and immediate life-expressions, to the use of radio, television, telephones, teleprinters, communications satellites, computers, and so forth. All of these change the material conditions of communication, and may indeed be referred to as 'material forces of communication'. Yet such material forces of communication cannot play the same role in relation to a realist theory of social science as the material forces of inquiry play in relation to a realist theory of natural science.

Communications media do not make possible greater penetration into an 'independent reality', as scientific instruments do; they do not facilitate greater interpretative 'depth', since it is not clear whether any such depth, by analogy to that achieved by scientific instruments, can be meaningfully conceptualized. This does not, of course, mean that such material forces of communication have no impact on the social and cultural sciences, only that they do not, as scientific instruments do, make their contribution by allowing penetration into 'inaccessible mechanisms' or 'unobservable, theoretical entities', and that their development cannot generate the primary experience of a change in scientific practice on which a realist theory could be based.

[47] By 'genuine' theories of ideology I mean those that do not use the word in such a catch-all sense that no contrast is made between that which is ideology and that which is not. Much modern usage in sociology uses the word to refer to all belief-systems, and thus has no theory of the specificity of ideology. Cf. Geuss (1981, pp. 4ff.).

[48] For further discussion of critical theory's conception of ideology, see Geuss (1981).

[49] In contrast, ethogenic social psychology has developed no coherent theory of ideology and may be ignored for present purposes.

[50] For a discussion of the concepts of 'internal' and 'external' goals of scientific research, see van den Daele *et al.* (1977); cf. Holzkamp (1972, pp. 17ff.). The question of the relationship between science policy and the theoretical development of sciences is a matter of much controversy, and it would not be appropriate to take it up further here. For a recent survey, see Rip (1981).

BIBLIOGRAPHY

Abbagano, N.: 1967, 'Positivism', in P. Edwards (ed.), *The Encyclopaedia of Philosophy*, Vol. 6, Macmillan, New York, pp. 414–419.

Abel, T.: 1948, 'The operation called Verstehen', *American Journal of Sociology* **54**, pp. 211–218.

Abrams, P., Deem, R., Finch, J., Rock, P. (eds.): 1981, *Practice and Progress: British Sociology 1950–1980*, George Allen and Unwin, London.

Achinstein, P. and Barker, S. F. (eds.): 1969, *The Legacy of Logical Positivism*, The John Hopkins Press, Baltimore.

Adorno, T. W., Albert, H., Dahrendorf, R., Habermas, J., Pilot, H., Popper, K. R.: 1969, *Der Positivismusstreit in der deutschen Soziologie*, Luchterhand, Neuwied.

Adorno, T. W., Albert, H., Dahrendorf, R., Habermas, J., Pilot, H., Popper, K. R.: 1976, *The Positivist Dispute in German Sociology*, translated by Glyn Adey and David Frisby, Heinemann Educational Books, London.

Ajdukiewicz, K.: 1973, *Problems and Theories of Philosophy*, translated by Henryk Skolimowski and Anthony Quinton, Cambridge University Press, Cambridge.

Albert, H.: 1968a, *Traktat über kritische Vernunft*, Mohr, Tübingen.

Albert, H.: 1968b, 'Review of J. Habermas, Zur Logik der Sozialwissenschaften', *Kölner Zeitschrift für Soziologie und Sozialpsychologie* **20**, pp. 341–345.

Albert, H.: 1971, *Plädoyer für kritischen Rationalismus*, Piper, Munich.

Albert, H.: 1975, *Transzendentale Träumereien*, Hoffmann und Campe, Hamburg.

Albert, H.: 1976a, 'Behind positivism's back?', in Adorno *et al.* (1976).

Albert, H.: 1976b, 'The myth of total reason', in Adorno *et al.* (1976).

Althusser, L.: 1969, *For Marx*, Penguin Books, Harmondsworth.

Althusser, L. and Balibar, E.: 1970, *Reading Capital*, New Left Books, London.

Apel, K.-O., v. Bormann, C., Bubner, R., Gadamer, H.-G., Giegel, H. J., Habermas, J.: 1971, *Hermeneutik und Ideologiekritik*, Suhrkamp, Frankfurt/M.

Apel, K.-O. (ed.): 1976, *Sprachpragmatik und Philosophie*, Suhrkamp, Frankfurt/M.

Apel, K.-O.: 1977, 'Types of social science in the light of human interests of knowledge', *Social Research* **44**, 425–470.

Armistead, N. (ed.): 1974, *Reconstructing Social Psychology*, Penguin Books, Harmondsworth.

Avenarius, R.: 1888–1890: *Kritik der reinen Erfahrung*, Reisland, Leipzig.

Ayer, A. J. (ed.): 1959, *Logical Positivism*, The Free Press, Glencoe, Illinois.

Ayer, A. J.: 1962, *Language, Truth and Logic*, Gollancz, London (first published 1936, second edition 1946).

Bauman, Z.: 1973, 'On the philosophical status of ethnomethodology', *Sociological Review* **21**, pp. 5–23.

Bauman, Z.: 1978, *Hermeneutics and Social Science*, Hutchinson, London.

Beck, L. W.: 1963, *A Commentary on Kant's Critique of Practical Reason*, University of Chicago Press, Chicago.

Benton, R.: 1977, *Philosophical Foundations of the Three Sociologies*, Routledge and Kegan Paul, London.

Benton, T.: 1981, 'Realism and social science: some comments on Roy Bhaskar's "The Possibility of Naturalism" ', *Radical Philosophy* **27**, pp. 13–21.

Berger, P. and Pullberg, S.: 1966, 'Reification and the sociological critique of consciousness', *New Left Review* **35**, pp. 57–71.

Berkowitz, L. (ed.): 1977, *Advances in Experimental Social Psychology*, Volume 10, Academic Press, New York.

Bhaskar, R.: 1975, 'Feyerabend and Bachelard: two philosophies of science', *New Left Review* **94**, pp. 31–55.

Bhaskar, R.: 1978a, *A Realist Theory of Science*, second edition, Harvester Press, Brighton.

Bhaskar, R.: 1978b, 'On the possibility of social scientific knowledge and the limits of naturalism', *Journal for the Theory of Social Behaviour* **8**, pp. 1–28.

Bhaskar, R.: 1979, *The Possibility of Naturalism*, Harvester Press, Brighton.

Bhaskar, R.: 1980, 'Scientific explanation and human emancipation', *Radical Philosophy* **26**, pp. 16–28.

Bleicher, J.: 1980, *Contemporary Hermeneutics*, Routledge and Kegan Paul, London.

Bloch, E.: 1971, *On Karl Marx*, Herder and Herder, New York.

Bohm, D.: 1964, 'On the problem of truth and understanding in science', in Bunge (ed.), (1964).

Bohm, D.: 1971, 'Fragmentation in science and society', in Fuller (ed.) (1971).

Böhme, G., van den Daele, W., and Krohn, W.: 1972, 'Alternativen in der Wissenschaft', *Zeitschrift für Soziologie* **1**, pp. 302–316.

Böhme, G., van den Daele, W., and Krohn, W.: 1975, 'Finalization in science', *Social Science Information* **15**, pp. 306–330.

Brody, B. A.: 1967, 'Natural kinds and real essences', *Journal of Philosophy* **64**, pp. 431–446.

Bukharin, N. I. *et al.*: 1971, *Science at the Crossroads*, Frank Cass, London.

Bunge, M. (ed.): 1964, *The Critical Approach to Science and Philosophy*, The Free Press of Glencoe, London.

Burtt, E. A.: 1925, *The Metaphysical Foundations of Modern Physical Science*, Kegan Paul, Trench, Trubner, London.

Campbell, D. T.: 1974, 'Evolutionary epistemology', in Schilpp (ed.), 1974.

Carnap, R.: 1935, *Philosophy and Logical Syntax*, Routledge and Kegan Paul, London.

Carnap, R.: 1959, 'The elimination of metaphysics through logical analysis of language', in Ayer (ed.) (1959).

Carnap, R.: 1967, *The Logical Structure of the World/Pseudoproblems in Philosophy*, University of California Press, Berkeley and Los Angeles.

Colletti, L.: 1972, *From Rousseau to Lenin*, New Left Books, London.

Comte, A.: 1875, *System of Positive Polity*, Longmans, Green, London.

Connerton, P. (ed.): 1976, *Critical Sociology*, Penguin Books, Harmondsworth.

Cornforth, M.: 1947/48, 'Reply to Sohn-Rethel', *Modern Quarterly* **3**, pp. 83–84.

Coulter, J.: 1971, 'Decontextualized meanings: current approaches to verstehende investigations', *The Sociological Review* **19**, pp. 301–323.

Dahrendorf, R.: 1976, 'Remarks on the discussion', in Adorno *et al.* (1976).

Denzin, N. K.: 1970, *The Research Act in Sociology*, Butterworths, London.

Dijksterhuis, E. J.: 1961, *The Mechanization of the World Picture*, Oxford University Press, Oxford.
Dilthey, W.: 1927, *Gesammelte Schriften*, Vol. VII, Teubner, Leipzig.
Douglas, J. D. (ed.): 1971, *Understanding Everyday Life*, Routledge and Kegan Paul, London.
Duhem, P.: 1954, *The Aim and Structure of Physical Theory*, translated by Philip P. Wiener, Princeton University Press, Princeton.
Durkheim, E.: 1961, *Moral Education*, The Free Press of Glencoe, New York.
Durkheim, E.: 1964a, *The Division of Labour in Society*, The Free Press of Glencoe, New York.
Durkheim, E.: 1964b, *The Rules of Sociological Method*, The Free Press of Glencoe, New York.
Durkheim, E.: 1974, *Sociology and Philosophy*, The Free Press, New York.
Emmet, D. and MacIntyre, A. (eds.): 1970, *Sociological Theory and Philosophical Analysis*, Macmillan, London.
Fahrenbach, H. (ed.): 1973, *Wirklichkeit und Reflexion*, Neske, Pfullingen.
Filmer, P., Phillipson, M., Silverman, D., Walsh, D.: 1972, *New Directions in Sociological Theory*, Collier-Macmillan, London.
Fisk, M.: 1973, *Nature and Necessity: An Essay in Physical Ontology*, Indiana University Press, Bloomington and London.
Fuller, W. (ed.): 1971, *The Social Impact of Modern Biology*, Routledge and Kegan Paul, London.
Gadamer, H.-G.: 1975, *Truth and Method*, Sheed and Ward, London.
Garfinkel, H.: 1967, *Studies in Ethnomethodology*, Prentice-Hall, Englewood Cliffs, N.J.
Gasking, D.: 1955, 'Causation and recipes', *Mind* **64**, pp. 479–487.
Geuss, R.: 1981, *The Idea of a Critical Theory: Habermas and the Frankfurt School*, Cambridge University Press, Cambridge.
Giddens, A. (ed.): 1974, *Positivism and Sociology*, Heinemann Educational Books, London.
Giddens, A.: 1976, *New Rules of Sociological Method*, Hutchinson, London.
Giddens, A.: 1977, *Studies in Social and Political Theory*, Hutchinson, London.
Giedymin, J.: 1975, 'Antipositivism in contemporary philosophy of social science and humanities', *British Journal for the Philosophy of Science* **26**, pp. 275–301.
Giegel, H. J.: 1971, 'Reflexion und Emanzipation', in Apel *et al.* (1971).
Glaser, B. G. and Strauss, A. L.: 1967, *The Discovery of Grounded Theory*, Weidenfeld and Nicolson, London.
Goffman, E.: 1969, *Where the Action Is*, Allen Lane, The Penguin Press, London.
Goldmann, L.: 1971, *Immanuel Kant*, New Left Books, London.
Goodman, N.: 1954, *Fact, Fiction and Forecast*, Athlone Press, London.
Grossmann, H.: 1935, 'Die gesellschaftlichen Grundlagen der mechanistischen Philosophie und die Manufaktur', *Zeitschrift für Sozialforschung* **4**, pp. 161–231.
Habermas, J.: 1966, 'Knowledge and interest', *Inquiry* **9**, pp. 285–300. Reprinted in Emmet and MacIntyre (eds.) (1970).
Habermas, J.: 1968, *Technik und Wissenschaft als 'Ideologie'*, Suhrkamp, Frankfurt/M.
Habermas, J.: 1970a, *Zur Logik der Sozialwissenschaften*, Suhrkamp, Frankfurt/M.
Habermas, J.: 1970b, 'Towards a theory of communicative competence', *Inquiry* **13**, pp. 360–375.

Habermas, J.: 1971a, 'Der Universalitätsanspruch der Hermeneutik', in Apel *et al.* (1971).

Habermas, J.: 1971b, *Toward a Rational Society*, Heinemann Educational Books, London.

Habermas, J.: 1972, *Knowledge and Human Interests*, Heinemann Educational Books, London.

Habermas, J.: 1973a, 'A postscript to Knowledge and Human Interests', *Philosophy of the Social Sciences* **3**, pp. 157–189.

Habermas, J.: 1973b, *Erkenntnis und Interesse*, Suhrkamp, Frankfurt/M.

Habermas, J.: 1973c, 'Wahrheitstheorien', in Fahrenbach (ed.) (1973).

Habermas, J.: 1974, *Theory and Practice*, translated by John Viertel, Heinemann Educational Books, London.

Habermas, J.: 1976a, 'Was heisst Universalpragmatik?', in Apel (ed.) (1976).

Habermas, J.: 1976b, *Legitimation Crisis*, Heinemann Educational Books, London.

Habermas, J. and Luhmann, N.: 1971, *Theorie der Gesellschaft oder Sozialtechnologie*? Suhrkamp, Frankfurt/M.

Halfpenny, P.: 1976, *Explanations in Sociology: Positivist and Interpretivist Models*, unpublished Ph.D. Thesis, University of Essex.

Harré, R.: 1961, *Theories and Things*, Sheed and Ward, London.

Harré, R.: 1970, *The Principles of Scientific Thinking*, Macmillan, London.

Harré, R.: 1972, *The Philosophies of Science*, Oxford University Press, London.

Harré, R.: 1974, 'Blueprint for a new science', in Armistead (ed.) (1974).

Harré, R.: 1977a, 'The ethogenic approach: theory and practice', in Berkowitz (ed.) (1977).

Harré, R.: 1977b, 'Automatisms and autonomies: in reply to Professor Schlenker', in Berkowitz (ed.) (1977).

Harré, R.: 1979, *Social Being: A Theory for Social Psychology*, Basil Blackwell, Oxford.

Harré, R. and Madden, E.: 1975, *Causal Powers*, Basil Blackwell, Oxford.

Harré, R. and Secord, P.: 1972, *The Explanation of Social Behaviour*, Basil Blackwell, Oxford.

Held, D.: 1980, *Introduction to Critical Theory: Horkheimer to Habermas*, Hutchinson, London.

Heller, A.: 1979, 'Can the Unity of Sciences be Considered as the Norm of Sciences', in Nowotny and Rose (eds.) (1979).

Hempel, C. G.: 1965, *Aspects of Scientific Explanation*, The Free Press, New York.

Hempel, C. G.: 1966, *Philosophy of Natural Science*, Prentice-Hall, Englewood Cliffs, N.J.

Henry, M.: 1976, *Marx*, Gallimard, Paris.

Hesse, M.: 1974, *The Structure of Scientific Inference*, Macmillan, London.

Hessen, B.: 1971, 'The social and economic roots of Newton's "Principia" ', in Bukharin *et al.* (1971).

Hirst, R. M.: 1967, 'Realism', in P. Edwards (ed.), *The Encyclopaedia of Philosophy*, Macmillan, New York, Vol. 7, pp. 177–83.

Hollis, M.: 1977, *Models of Man*, Cambridge University Press, Cambridge.

Holzkamp, K.: 1972, *Kritische Psychologie*, Fischer, Frankfurt/M.

Hooker, C. A.: 1974, 'Systematic realism', *Synthese* **26**, pp. 409–497.

Hooker, C. A.: 1975, 'Philosophy and meta-philosophy: empiricism, Popperianism, and realism', *Synthese* **32**, pp. 177–231.

Horkheimer, M.: 1967, 'Ein neuer Ideologiebegriff?' in Lenk (ed.) (1967).
Horkheimer, M.: 1972, *Critical Theory: Selected Essays*, translated by M. J. O. O'Connell and others, Herder and Herder, New York.
Horkheimer, M. and Adorno, T. W.: 1973, *Dialectic of Enlightenment*, translated by John Cumming, Allen Lane, The Penguin Press, London.
Hüber, K., Lobkowicz, N., Lübbe, H., Radnitzky, G. (eds.): 1976, *Die politische Herausforderung der Wissenschaft*, Hoffmann und Campe, Hamburg.
Hume, D.: 1902, *Enquiries Concerning the Human Understanding and Concerning the Principles of Morals*, edited by L. A. Selby-Bigge, The Clarendon Press, Oxford.
Iggers, G. G.: 1968, *The German Conception of History*, Wesleyan University Press, Middletown.
Jay, M.: 1973, *The Dialectical Imagination*, Heinemann, London.
Johansson, I.: 1975, *A Critique of Karl Popper's Methodology*, Akademiförlaget, Stockholm.
Kant, I.: 1929, *Critique of Pure Reason*, translated by Norman Kemp Smith, Macmillan, London.
Kant, I.: 1953, *Prolegomena*, translated by Peter G. Lucas, Manchester University Press, Manchester.
Kant, I.: 1963, *On History*, edited by Lewis White Beck, Bobbs-Merrill, Indianapolis.
Keat, R.: 1971, 'Positivism, naturalism, and anti-naturalism in the social sciences', *Journal for the Theory of Social Behaviour* 1, pp. 3–17.
Keat, R.: 1981, *The Politics of Social Theory: Habermas, Freud and the Critique of Positivism*, Basil Blackwell, Oxford.
Keat, R. and Urry, J.: 1975, *Social Theory as Science*, Routledge and Kegan Paul, London.
Kołakowski, L.: 1972, *Positivist Philosophy: From Hume to the Vienna Circle*, Penguin Books, Harmondsworth.
Korsch, K.: 1970, *Marxism and Philosophy*, New Left Books, London.
Kortian, G.: 1980, *Metacritique: The Philosophical Argument of Jürgen Habermas*, Cambridge University Press, Cambridge.
Koyré, A.: 1968, *Metaphysics and Measurement*, Chapman and Hall, London.
Kraft, V.: 1953, *The Vienna Circle: The Origin of Neo-Positivism*, Philosophical Library, New York.
Kreckel, R.: 1972, *Soziologische Erkenntnis und Geschichte*, Westdeutscher Verlag, Opladen.
Kreckel, R.: 1975, *Soziologisches Denken: eine kritische Einführung*, Leske Verlag, Opladen.
Kuhn, T.: 1962, *The Structure of Scientific Revolutions*, University of Chicago Press, Chicago and London.
Lakatos, I. and Musgrave, A. (eds.): 1970, *Criticism and the Growth of Knowledge*, Cambridge University Press, Cambridge.
Leat, D.: 1972, 'Misunderstanding Verstehen', *The Sociological Review* **20**, pp. 29–38.
Lecourt, D.: 1975, *Marxism and Epistemology*, New Left Books, London.
Leiss, W.: 1972, 'Technological rationality: Marcuse and his critics', *Philosophy of the Social Sciences* 2, pp. 31–42.
Leiss, W.: 1975, 'The problem of man and nature in the work of the Frankfurt School', *Philosophy of the Social Sciences* 5, pp. 163–172.

Lenk, K. (ed.): 1967, *Ideologie*, Luchterhand, Neuwied and Berlin (third edition).
Lindesmith, A. R. and Strauss, A. L.: 1968, *Social Psychology*, Rinehart and Winston, New York.
Lobkowicz, N.: 1972, 'Interest and objectivity', *Philosophy of the Social Sciences* 2, pp. 193–210.
Locke, J.: 1959, *An Essay Concerning Human Understanding*, Dover, New York.
Losee, J.: 1972, *A Historical Introduction to the Philosophy of Science*, Oxford University Press, London.
Lukács, G.: 1971, *History and Class Consciousness*, Merlin Press, London.
Madden, E. H. and Sachs, M.: 1972, 'Parmenidean particulars and vanishing elements', *Studies in History and Philosophy of Science* 3, pp. 151–166.
Manis, J. G. and Meltzer, B. N. (eds.): 1972, *Symbolic Interactionism: A Reader in Social Psychology*, Allyn and Bacon, Boston.
Marcuse, H.: 1955, *Reason and Revolution*, Routledge and Kegan Paul, London.
Marcuse, H.: 1967, 'Über das Ideologieproblem in der hochentwickelten Industriegesellschaft', in Lenk (ed.) (1967).
Marcuse, H.: 1968, *One-Dimensional Man*, Sphere Books, London.
Marcuse, H.: 1971, *Soviet Marxism*, Penguin Books, Harmondsworth.
Marcuse, H.: 1972a, *Negations: Essays in Critical Theory*, Penguin Books, Harmondsworth.
Marcuse, H.: 1972b, *Counterrevolution and Revolt*, Beacon Press, Boston.
Marsh, P., Rosser, E. and Harré, R.: 1978, *The Rules of Disorder*, Routledge and Kegan Paul, London.
Marx, K.: 1970, *A Contribution to the Critique of Political Economy*, Progress Publishers, Moscow.
Marx, K.: 1973, *Grundrisse*, Penguin Books, Harmondsworth.
Marx, K.: 1975, *Early Writings*, edited by Lucio Colletti, Penguin Books, Harmondsworth.
Marx, K.: 1976, *Capital*, Vol. 1, Penguin Books, Harmondsworth.
Marx, K.: 1981, *Capital*, Vol. 3, Penguin Books, Harmondsworth.
Marx, K. and Engels, F.: 1968, *The German Ideology*, Progress Publishers, Moscow.
Marx, K. and Engels, F.: 1973, *Werke*, Dietz Verlag, Berlin.
Medawar, P. B.: 1969, *Induction and Intuition in Scientific Thought*, Methuen, London.
Mendelsohn, E., Weingart, P. and Whitley, R. (eds.): 1977, *The Social Production of Scientific Knowledge*, D. Reidel, Dordrecht.
Meszaros, I.: 1970, *Marx's Theory of Alienation*, Merlin Press, London.
Mill, S.: 1961, *Auguste Comte and Positivism*, Ann Arbor Paperbacks, University of Michigan Press, Ann Arbor (first published 1865).
Miller, D.: 1972, 'Back to Aristotle?' *British Journal for the Philosophy of Science* 23, pp. 69–79.
Milligan, M. O.: 1948/49, 'Karl Popper's positivism and the science of society', *The Modern Quarterly* 4, pp. 51–66.
Misgeld, D.: 1977, 'Critical theory and hermeneutics: the debate between Habermas and Gadamer', in O'Neill (ed.) (1977).
Montefiore, A.: 1976, 'The French philosophical tradition', *The Listener* 95:2459, 27.5.1976.
Münch, P.: 1973, *Gesellschaftstheorie und Ideologiekritik*, Hoffmann und Campe, Hamburg.

McCarthy, T. A.: 1972, 'A theory of communicative competence', *Philosophy of the Social Sciences* 3, 135–156.
McCarthy, T.: 1978, *The Critical Theory of Jürgen Habermas*, Hutchinson, London.
McHugh, P., Raffel, S., Foss, D. C., Blum, A. F.: 1974, *On the Beginnings of Social Inquiry*, Routledge and Kegan Paul, London.
Macpherson, C. B.: 1966, *The Real World of Democracy*, The Clarendon Press, Oxford.
Nagel, E.: 1961, *The Structure of Science*, Routledge and Kegan Paul, London.
Needham, J.: 1969, *The Grand Titration*, Allen and Unwin, London.
Neurath, O., Carnap, R. and Morris, C., (eds.): 1970, *Foundations of the Unity of Science*, University of Chicago Press, London and Chicago.
Nowotny, H. and Rose, H. (eds.): 1979, *Counter-Movements in the Sciences*, D. Reidel, Dordrecht.
Ollman, B.: 1971, *Alienation*, Cambridge University Press, Cambridge.
O'Neill, J. (ed.): 1977, *On Critical Theory*, Heinemann, London.
Outhwaite, W.: 1975, *Understanding Social Life: The Method called Verstehen*, Allen and Unwin, London.
Palmer, R. E.: 1969, *Hermeneutics*, Northwestern University Press, Evanston.
Parsons, T.: 1949, *The Structure of Social Action*, The Free Press of Glencoe, New York.
Parsons, T.: 1971, 'Value-freedom and objectivity', in Stammer (ed.) (1971).
Passmore, J.: 1968, *A Hundred Years of Philosophy*, Penguin Books, Harmondsworth.
Pears, D.: 1971, *Wittgenstein*, Fontana/Collins, London.
Peirce, C. S.: 1931–1935, 1958, *Collected Papers*, edited by Charles Hartshorne and Paul Weiss (Vols. I–VI) and by Arthur W. Burks (Vols. VII–VIII) Harvard University Press, Cambridge, Mass.
Pfetsch, F. R.: 1979, 'The "finalization" debate in Germany: some comments and explanations', *Social Studies of Science* 9, 115–124.
Pilot, H.: 1976, 'Jürgen Habermas' empirically falsifiable philosophy of history', in Adorno *et al.* (1976).
Platt, J.: 1981, 'The social construction of "positivism" and its significance in British sociology, 1950–1980', in Abrams *et al.* (eds.) (1981).
Popper, K.: 1961, *The Poverty of Historicism*, Routledge and Kegan Paul, London.
Popper, K.: 1962, *The Open Society and its Enemies*, Routledge and Kegan Paul, London (fourth edition).
Popper, K.: 1968, *The Logic of Scientific Discovery*, Hutchinson, London.
Popper, K.: 1969, *Conjectures and Refutations*, Routledge and Kegan Paul, London (third edition).
Popper, K.: 1972, *Objective Knowledge: An Evolutionary Approach*, The Clarendon Press, Oxford.
Popper, K.: 1976, *Unended Quest: An Intellectual Autobiography*, Fontana/Collins, London.
Putnam, H.: 1978, *Meaning and the Moral Sciences*, Routledge and Kegan Paul, London.
Quine, W. v. O.: 1961, *From a Logical Point of View*, Harvard University Press, Cambridge, Mass.
Reichenbach, H.: 1951, *The Rise of Scientific Philosophy*, University of California Press, Berkeley.
Rescher, N.: 1977, *Methodological Pragmatism*, Basil Blackwell, Oxford.

Rickman, H. P.: 1976, *W. Dilthey: Selected Writings*, Cambridge University Press, Cambridge.
Rip, A.: 1981, 'A cognitive approach to science policy', *Research Policy* **10**, pp. 294–311.
Ruben, D.-H.: 1977, *Marxism and Materialism*, Harvester Press, Sussex.
Runciman, W. G.: 1963, *Social Science and Political Theory*, Cambridge University Press, Cambridge.
Ryan, A.: 1970, *The Philosophy of the Social Sciences*, Macmillan, London.
Ryle, G.: 1963, *The Concept of Mind*, Penguin Books, Harmondsworth.
Schillp, P. A. (ed.): 1974, *The Philosophy of Karl Popper* (2 Vols.), Open Court Publishing Co., La Salle.
Schlick, M.: 1959, 'Positivism and realism', in Ayer (ed.) (1959).
Schmidt, A. (ed.): 1969, *Beiträge zur marxistischen Erkenntnistheorie*, Suhrkamp, Frankfurt/M.
Schmidt, A.: 1971, *The Concept of Nature in Marx*, New Left Books, London.
Schnädelbach, H.: 1969, 'Was ist Ideologie?' *Das Argument* **50**, pp. 71–92.
Schnädelbach, H.: 1971, *Erfahrung, Begründung, und Reflexion: Versuch über den Positivismus*, Suhrkamp, Frankfurt/M.
Schnädelbach, H.: 1972, 'Über den Realismus', *Zeitschrift für allgemeine Wissenschaftstheorie* **3**, pp. 88–112.
Scott, J. P.: 1978, 'Critical social theory: an introduction and critique', *British Journal of Sociology* **29**, pp. 1–21.
Simon, W. M.: 1972, *European Positivism in the Nineteenth Century*, Cornell University Press/Kennikat Press, New York.
Sklair, L.: 1970, *The Sociology of Progress*, Routledge and Kegan Paul, London.
Slater, P.: 1977, *The Origin and Significance of the Frankfurt School*, Routledge and Kegan Paul, London.
Sohn-Rethel, A.: 1947/48, 'Materialism and its advocacy', *Modern Quarterly* **3**, pp. 74–83.
Sohn-Rethel, A.: 1972, *Geistige und körperliche Arbeit*, Suhrkamp, Frankfurt/M.
Sohn-Rethel, A.: 1978, *Intellectual and Manual Labour: A Critique of Epistemology*, Macmillan, London.
Der Spiegel: 1976, 30/38, 13 September.
Stammer, O. (ed.): 1971, *Max Weber and Sociology Today*, Basil Blackwell, Oxford.
Stevenson, C. L.: 1944, *Ethics and Language*, Yale University Press, New Haven and London.
Stockman, N.: 1978, 'Habermas, Marcuse, and the *Aufhebung* of Science and Technology', *Philosophy of the Social Sciences* **8**, pp. 15–35.
Suppe, F. (ed.): 1977, *The Structure of Scientific Theories*, University of Illinois Press, Urbana (second edition).
Taylor, C.: 1964, *The Explanation of Behaviour*, Routledge and Kegan Paul, London.
Thompson, J. B.: 1981, *Critical Hermeneutics: A Study in the Thought of Paul Ricoeur and Jürgen Habermas*, Cambridge University Press, Cambridge.
van den Daele, W., Krohn, W. and Weingart, P.: 1977, 'The political direction of scientific development', in Mendelsohn *et al.* (eds.) (1977).
Wax, M. L.: 1966/67, 'On misunderstanding Verstehen: a reply to Abel', *Sociology and Social Research* **51**, pp. 323–333.

Wellmer, A.: 1967, *Methodologie als Erkenntnistheorie*, Suhrkamp, Frankfurt/M.
Wellmer, A.: 1969, *Kritische Gesellschaftstheorie und Positivismus*, Suhrkamp, Frankfurt/M.
Wellmer, A.: 1971, *Critical Theory of Society*, translated by John Cumming, Herder and Herder, New York.
Wellmer, A.: 1977, 'Communications and emancipation: reflections on the linguistic turn in critical theory', in O'Neill (ed.) (1977).
Williams, R.: 1976, *Keywords: A Vocabulary of Culture and Society*, Fontana, London.
Winch, P.: 1958, *The Idea of a Social Science*, Routledge and Kegan Paul, London.
Wittgenstein, L.: 1961, *Tractatus Logico-philosophicus*, translated by D. F. Pears and B. F. McGuiness, Routledge and Kegan Paul, London.
Wittgenstein, L.: 1963, *Philosophical Investigations*, Basil Blackwell, Oxford.
Woozley, A. D.: 1967, 'Universals', in P. Edwards (ed.), *The Encyclopaedia of Philosophy*, Macmillan, New York, Vol. 8, pp. 194–220.
von Wright, G. H.: 1971, *Explanation and Understanding*, Routledge and Kegan Paul, London.
Zilsel, E.: 1942a, 'The sociological roots of science', *American Journal of Sociology* **47**, pp. 544–562.
Zilsel, E.: 1942b, 'The genesis of the concept of physical law', *The Philosophical Review* **51**, pp. 245–279.

INDEX

MIX
Papier aus verantwortungsvollen Quellen
Paper from responsible sources
FSC® C105338

If you have any concerns about our products,
you can contact us on
ProductSafety@springernature.com

In case Publisher is established outside the EU,
the EU authorized representative is:
Springer Nature Customer Service Center GmbH
Europaplatz 3, 69115 Heidelberg, Germany

Printed by Libri Plureos GmbH
in Hamburg, Germany